Group Invariance in Engineering
Boundary Value Problems

R. Seshadri · T.Y. Na

Group Invariance in Engineering Boundary Value Problems

With 25 Figures

Springer-Verlag
New York Berlin Heidelberg Tokyo

R. Seshadri
Syncrude Canada Limited
Fort McMurray, Alberta
Canada T9H 3LI

T.Y. Na
Department of Mechanical
Engineering
University of Michigan—Dearborn
Dearborn, Michigan 48128
U.S.A.

Library of Congress Cataloging in Publication Data
Seshadri, R.
Group invariance in engineering boundary value problems.
Includes bibliographies and index.
1. Boundary value problems. 2. Transformation groups.
I. Na, T.Y. II. Title.
TA347.B69S47 1985 515.3'5 84-26886

Printed and bound by R.R. Donnelley & Sons, Harrisonburg, Virginia.
Printed in the United States of America.

9 8 7 6 5 4 3 2 1

ISBN 0-387-96128-3 Springer-Verlag New York Berlin Heidelberg Tokyo
ISBN 3-540-96128-3 Springer-Verlag Berlin Heidelberg New York Tokyo

Preface

In the latter part of the last century, Sophus Lie first introduced and developed quite extensively the theory of continuous groups of transformations in connection with the study of differential equations. In the last few decades, there has been a revival of interest in group-theoretic methods and significant progress has been made due to the efforts of several mathematicians, engineers and physicists.

Group-theoretic methods are powerful, versatile and fundamental to the development of systematic procedures that lead to invariant solutions of boundary value problems. Since the group methods are not based on linear operators, superposition or other requirements of the linear solution techniques, they are applicable to both linear and nonlinear differential models. A number of books on the application of the continuous groups of transformations relating to differential equations have been written from a mathematical standpoint. In dealing with differential boundary value problems in engineering and applied science however, physical aspects associated with the problems are of importance. Consideration of boundary and initial conditions as an integral part of the mathematical description becomes an essential part of any group-theoretic analysis. The purpose of this book is to provide a comprehensive and systematic treatment of group-theoretic methods from a standpoint of engineering and applied science, with particular emphasis on boundary value problems. The book is intended for senior undergraduate students, graduate students and research workers in the areas of engineering and applied science.

The authors are indebted to Arthur Na for his assistance during the preparation of the manuscripts. The second author wishes to express his sincere gratitude to Dr. Arthur G. Hansen for introducing him to this important method of analysis and for the advice and encouragement received through the years. The authors would like to dedicate this book to Dr. Arthur G. Hansen in view of the key contributions made by him to the area of similarity analysis pertaining to engineering boundary value problems.

CONTENTS

Chapter 1

INTRODUCTION AND GENERAL OUTLINE

Physical problems in engineering science are often described by differential models either linear or nonlinear. There is also an abundance of transformations of various types that appear in the literature of engineering and mathematics that are generally aimed at obtaining some sort of simplification of a differential model.

Similarity transformations [1,2] * which essentially reduce the number of independent variables in partial differential systems, have been widely used in fluid mechanics and heat transfer. Transformations have also been used to convert a boundary value problem to an initial value problem suitable for numerical procedures that employ forward marching schemes [3]. In other instances transformations have been used to reduce the order of an ordinary differential equation. Mappings have also been discovered which transform nonlinear partial or ordinary differential equations to linear forms. Underlying these seemingly unrelated transformations is a unified general principle which is based on the theory of continuous group of transformations. The theory was first introduced and developed extensively by Sophus Lie in the latter part of the last century. In recent years, there has been a revival of interest in applying the principles of continuous group of transformations to differential models, linear as well as nonlinear. In 1950, Birkhoff [4] proposed a method based upon simple groups of transformations for obtaining invariant solutions for some problems in the general area of hydrodynamics. The method essentially involves algebraic manipulations, an aspect which makes the method attractive. Group - theoretic methods are a powerful tool because they are not based on linear operators, superposition or any other aspects of linear solution techniques. Therefore, these methods are applicable to nonlinear differential models.

A majority of the recent books on the applications of continuous transformation groups [5,6] have been approached from a mathematical standpoint. In dealing with different boundary value problems in engineering science, the physical aspects associated with the problem need to be properly addressed. The treatment of boundary conditions as an integral part of the differential model in group - theoretic methods becomes relevant. The purpose of this book is to provide a comprehensive treatment of the subject from a standpoint of engineering science, with special reference to boundary value problems. Applications of the group-theoretic principles involved are presented in a clear and systematic fashion. The contents and the treatment of this book are particularly suitable for senior undergraduate students, graduate students and analytical workers in the area of engineering and applied sciences.

---

* Numbers in superscripts refer to numbers of references at the end of the chapter.

The concepts of continuous transformation groups are presented in chapter two. This chapter contains the theoretical background needed for all the subsequent chapters, and forms the basis for the entire book.

Chapter three contains a survey of the available methods for determining similarity transformations. A familiar form of the similarity transformation is given by $u(x,y) = x^{\alpha} f(xy^{\beta})$, where α and β take on values such that the original partial differential equation with independent variables x, y and dependent variable u is transformed to an ordinary differential equation with variables f and xy^{β}. The reduction in the number of independent variables is a simplification of the original mathematical description whether the resulting description is solved analytically, numerically or by using other approximate procedures. Consider, as an example, the one-dimensional nonlinear diffusion equation

$$\frac{\partial}{\partial x}[D(u)\frac{\partial u}{\partial x}] = \frac{\partial u}{\partial t} \tag{1.1}$$

A similarity transformation of the form

$$u = F(\varsigma) \quad ; \quad \varsigma = \frac{x}{2\sqrt{t}} \tag{1.2}$$

would transform equation (1.1) to an ordinary differential equation of the form

$$\frac{d}{d\varsigma}[D(F)\frac{dF}{d\varsigma}] + 2\varsigma\frac{dF}{d\varsigma} = 0 \tag{1.3}$$

The similarity transformation, equation (1.2), is just one type of transformation that can be obtained by direct procedures based on assumed transformation of the type $u(x,t) = t^{\alpha} F(x/t^{\delta})$, or by the use of dimensional group of transformations. The use of deductive group procedures which start out with a general group of transformations lead to some similarity solutions that are not obtainable by inspectional group procedures. The group - theoretic methods of similarity analysis imply that the search for similarity solutions of a system of partial differential equations is equivalent to the determination of solutions of these equations invariant under a group of transformations. For boundary value problems, it follows that the auxiliary conditions also be invariant under the same group of transformations. In chapter three, the methods for the determination of similarity transformations are classified into (i) direct methods and (ii) group - theoretic methods. The group - theoretic methods are further divided into inspectional and deductive procedures. In chapter four, the direct as well as group - theoretic methods are applied to a variety of nonlinear boundary value problems arising in engineering science.

Traditionally, similarity solutions have been discovered for boundary value problems with semi - infinite or infinite domains thus restricting the solutions to a narrow class of problems. Chapter five examines the applicability of similarity analysis to boundary value problems in finite domains.

The construction of non - similar solutions from similarity solutions enables one to extend the similarity methods to a larger class of boundary value problems. Non - similar solutions are a result of either the equations or boundary conditions not being invariant under a given group of transformations. Techniques such as superposition of similarity solutions, fundamental solutions and pseudo - similarity analysis are discussed in chapter six for the purpose of obtaining non - similar solutions.

For moving boundary problems in general, it is necessary to locate the similarity coordinate at the moving boundary. When the governing partial differential equations are parabolic it should also be ascertained whether the speed of propagation of the moving boundary is finite or infinite. Typically, analysis of problems that involve a change of phase can be associated with a moving boundary that propagates at a finite speed. However if no phase change is involved, the propagation of the boundary would be either at a finite or an infinite speed. Chapter seven contains a detailed discussion of these ideas through appropriate examples.

Chapter eight deals with boundary value problems that involve propagation of waves. When the propagation of disturbances are along the characteristics of the equations, use is made of the similarity - characteristic (SC) relationship for determining the similarity coordinate at the wave front. In other instances, where propagation of waves is not necessarily along the characteristics such as shock waves, dispersive and traveling waves, the role of group invariance in obtaining similarity solutions is discussed.

The technique for transforming boundary value problems to initial value problems is discussed in chapter nine. Two methods are described : (1) the inspectional group method, and (2) the infinitesimal group method. Both methods start out by defining a group of transformations. The particular transformation within this group of transformations which can reduce the boundary value problem to an initial value problem is identified by stipulating the requirement that (a) the given differential equation be independent of the parameter of the transformation, and (b) the missing boundary condition be identified with the parameter of transformation.

Inspectional as well as deductive procedures for transforming a nonlinear differential equation to a linear differential equation is discussed in chapter ten. These procedures are based on the use of groups of point transformations which act on a finite dimensional space. Effort has been made to keep the treatment of the subject as simple as possible and to bring out the underlying principles involved clearly. Recent developments have shown that differential equations can be invariant under a continuous group of transformations beyond point or contact transformations. The groups known as the Lie - Bäcklund (LB) transformations act on an infinite dimensional space. Details for discovering mappings that transform nonlinear to linear differential equations using LB transformations are available in the works of Bluman and Kumei [7],and Anderson et al [8] . The LB transformation approach is not covered in this book.

Miscellaneous topics such as the reduction of differential equations into algebraic equations, reduction of the order of an ordinary differential equation, transformation of ordinary to partial differential equations, and reduction of the number of variables using multiparameter group of transformations are covered in some detail in chapter eleven. Also covered in this chapter are self - similar solutions of the first and second kind [9,10]. The Hellums - Churchill procedure for the determination of normalized representation suitable for dimensional scale - modeling and semi - analytical investigations is briefly discussed.

Throughout the entire book, the authors have endeavoured to cater to the needs of (a) a novice in the area of group theory, and (b) the analytical worker seeking to use the more rigorous deductive group procedures. It is hoped that the book will adequately address some of the needs of students and researchers in engineering science who are seeking to apply group - theoretic methods to nonlinear boundary value problems.

REFERENCES

[1] Hansen, A. G., Similarity Analysis of Boundary Value Problems in Engineering, Prentice Hall (1964).

[2] Ames, W. F., Nonlinear Partial Differential Equations in Engineering, Academic Press (1965).

[3] Na, T. Y., Computational Methods in Engineering Boundary Value Problems, Academic Press (1979).

[4] Birkhoff, G., Hydrodynamics, Princeton University Press (1950).

[5] Bluman, G. W. and Cole, J. D., Similarity Methods for Differential Equations, Springer - Verlag, New York (1974).

[6] Ovsiannikov, L. V., Group Analysis in Differential Equations, Academic Press (English translation edited by W. F. Ames, 1982).

[7] Bluman, G. W. and Kumei, S., "On the Remarkable Nonlinear Diffusion Equation," J. Math. Phys., Vol. 21, No. 5, (1980).

[8] Anderson, R. L., Kumei, S. and Wulfman, C. E., Rev. Mex. Fis. 21, 1, 35 (1972); Phys. Rev. Lett. 28, 988 (1972).

[9] Sedov, L. I., Similarity and Dimensional Method in Mechanics, Academic Press (1959).

[10] Zel'dovich, Ya. B. and Raizer, Yu. P., Physics of Shock Waves and High Temperature Hydrodynamic Phenomena, Hayes W. D. and Probstein, R. F. (Editors), Vol. 2, Academic Press (1967).

Chapter 2

CONCEPTS OF CONTINUOUS TRANSFORMATION GROUPS

2.0 Introduction

The foundation of the group - theoretic method is contained in the general theories of continuous transformation groups that were introduced and treated extensively by Lie [1], Lie and Engel [2] and Lie and Scheffers [3] in the latter part of the last century. Subsequently, the books by Cohen [4], Campbell [5], Eisenhart [6], Ovsjannikov [7], Bluman and Cole [8], have contributed greatly to the development and clarification of many of Lie's theories, particularly its applications to the invariant solutions of differential equations. In the literature of engineering and applied sciences, the works of Birkhoff [9], Morgan [10], Hansen [11], Na, Abbott and Hansen [12], Na and Hansen [13] and Ames [14.15] give quite extensively the general theories involved in the similarity solutions of partial differential equations as applied to engineering problems. It is assumed, however, that the average engineer may not be thoroughly acquainted with the concepts of that branch of modern algebra designated as group theory. For this reason as well as for clarifications of the terms and concepts involved, a brief review of some of the key aspects of the theory of transformation groups will be given in this chapter. Emphasis will be placed on presenting the ideas of Lie groups in a simple and clear manner suitable for an engineer and scientist, instead of the rigorous and mathematically elegant approach used in the books by Eisenhart [6], Ovsjannikov [7] and Bluman and Cole [8].

2.1 Group Approach

Quite simply, an algebraic group is a set (collection of elements) which has some sort of operation defined between its elements. In addition, a certain set of rules and statements regarding the elements and the defined operation must be satisfied. The elements in a set can be anything: integers, complex numbers, vectors, matrices, transformations etc. One important criterion, however, is the definition of an operation of these elements. Typical operations are integer additions, complex number multiplications, vector additions and successive transformations.

The rules which a set of elements must obey under a given operation are stated below, where the symbol $*$ denotes the binary operation between two elements a and b of a set G. A set G is called a group if

(1) The set of elements is closed under a given operation. If a and b are two elements of the set, then

$$a*b = c$$

where c is also an element of the set.

(2) There exists an identity element I such that

$$a*I = I*a = a$$

(3) Every element in G has an inverse in G for the operation $*$. Thus, given any element a, there exists an element a^{-1}, known as its inverse such that

$$a*a^{-1} = a^{-1}*a = I$$

(4) The operation $*$ is associative. Thus,

$$a*(b*c) = (a*b)*c$$

As an example of a group, let us consider the set of integers Z associated with the operation of addition in Z.

(1) Since addition of an integer to an integer gives an integer, the set Z is closed.

(2) If n is an integer, then

$$n + 0 = 0 + n = n$$

and the set consists of an identity element (0).

(3) Since the set Z includes both positive and negative integers, the set has the property that

$$n + (-n) = (-n) + n = 0$$

(4) If l, m and n are elements of the set Z then

$$(l + m) + n = l + (m + n)$$

Therefore, the set Z associated with the operation of addition is a group. This group is usually referred to as the additive group.

As a second example, consider the multiplicative group of non-zero complex numbers. Let C be the set of all non-zero complex numbers associated with the operation of multiplication whose elements are x = a + ib, where a and b are real numbers, and x $\neq$ 0 + i0. The following properties can be established:

(1) The set is closed, since the product

$$(a + ib)(c + id) = (ac - bd) + i(ad + bc)$$

is a unique element in C.

(2) The element 1 + i0 is clearly the identity element in C, since

$$(a + ib)(1 + i\,0) = (1 + i\,0)(a + ib) = a + ib$$

(3) The inverse of a given element a + ib is

$$\frac{a}{a^2+b^2} - i\frac{b}{a^2+b^2}$$

since

$$\left(\frac{a}{a^2+b^2} - i\frac{b}{a^2+b^2}\right)(a+ib) = (a+ib)\left(\frac{a}{a^2+b^2} - \frac{ib}{a^2+b^2}\right) = 1$$

(4) Suppose a + ib, c + id and e + if are elements of the set C, then

$$[(a+ib)(c+id)](e+if) = [(ac-bd)e - (bc+ad)f]$$
$$+i[(bc+ad)e + (ac-bd)f]$$
$$(a+ib)[(c+id)(e+if)] = [a(ce-df) - b(de+cf)]$$
$$+i[b(ce-df) + a(de+cf)]$$

It follows from above that the associative law is satisfied. As a result, the set of non-zero complex numbers associated with the operation of multiplication is a group.

2.2 Transformation Groups

Let $f^i(x^1, ..., x^n; a^1, ..., a^r)$; $(i = 1, ..., n)$ be a set of functions continuous in both the variables x^i and a^j. We will also assume the continuity of derivatives as may be required in the following discussions. The variables a^j are the parameters of the functions. The a^j are assumed to be "essential parameters", i.e., it is not possible to find (r - 1) functions of a^j: $\alpha^1(a)$, ..., $\alpha^{r-1}(a)$ such that

$$f^i(x^1, ..., x^n; a^1, ..., a^r) = F^i(x^1, ..., x^n; \alpha^1, ..., \alpha^{r-1})$$

If the a^j are not essential parameters, it means that fewer parameters can be constructed from the a^j to serve the same purpose in a function.

We now consider a set of functions f^i as a set of transformations depending on the parameters a^1, ..., a^r, and transforming a point $(x^1, ..., x^m)$ into $(\bar{x}^1, ..., \bar{x}^m)$, i.e.,

$$\bar{x}^i = f^i(x^1, ..., x^n; a^1, ..., a^r)$$

Successive transformations employing various sets of functions are considered to be the "operation" of the set. We will establish that the set of functions associated with the "operation" of a transformation is a group. Clearly, for this to be valid, the four conditions listed previously must be met. The conditions are

1. The set is closed.
2. There exists an identity transformation such that *

$$IT_{a1}x = T_{a1}Ix = T_{a1}x$$

3. Given any transformation T_{a1}, an inverse transformation ! exist such that

$$T_{a1}^{-1}T_{a1}x = T_{a1}T_{a1}^{-1}x = Ix = x$$

4. The associative law is true, i.e.,

$$T_{a1}(T_{a2}T_{a3})x = (T_{a1}T_{a2})T_{a3}x$$

As an example, let us consider the set of one-parameter transformation defined by

$$T_a x: \quad \bar{x}^1 = [x^1 cos(a) - x^2 sin(a)]e^a$$
$$\bar{x}^2 = [x^1 sin(a) + x^2 cos(a)]e^a$$

and prove that it constitutes a group of transformations. The parameter a is assumed to be real.

1. To show closure,· we have

$$T_{a2}T_{a1}x: \quad \bar{\bar{x}}^1 = [\bar{x}^1 cos(a^2) - \bar{x}^2 sin(a^2)]e^{a^2}$$
$$\bar{\bar{x}}^2 = [\bar{x}^1 sin(a^2) + \bar{x}^2 cos(a^2)]e^{a^2}$$

or,

$$\bar{\bar{x}}^1 = [(x^1 cos(a^1) - x^2 sin(a^1))cos(a^2)$$
$$-(x^1 sin(a^1) + x^2 cos(a^1))sin(a^2)]e^{a^2+a^1}$$
$$= [x^1 cos(a^1)cos(a^2) - x^2 sin(a^1)cos(a^2)$$
$$x^1 sin(a^1)sin(a^2) + x^2 cos(a^1)sin(a^2)]e^{a^2+a^1}$$

---

* For a particular set of a^j, say, $a_1^1, \ldots \ldots, a_1^r$, the transformations

$$\bar{x}^i = f(x^1, \ldots\ldots, a^n : a_1^1, \ldots\ldots, a_1^r)$$

is written as $T_{a1}x = \bar{x}$. Similarly, the function f^{*i} which represents the inverse functions of f^i, and transforms a point $(\bar{x}^1, \ldots\ldots, \bar{x}^n)$ into $(x^1, \ldots\ldots, x^n)$ can be written as $T_{a1}^{-1}\bar{x}$ =x. Two different sets of values are defined by different sets of values of the a^j. Thus, if $a_2^1, \ldots\ldots, a_2^r$ is a set of values distinct from $a_1^1, \ldots\ldots, a_1^r$, we consider T_{a1} and T_{a2} to be different transformations.

! For an inverse to exist, the Jacobian of the f^i with respect to the x^i is non-vanishing for a set of values of x^i.

$$= [(x^1 cos(a^1 + a^2) - x^2 sin(a^1 + a^2)]e^{a^1+a^2}$$

If $a^3{=}a^1{+}a^2$ is real, then we see that

$$\bar{\bar{x}}^1 = [x^1 cos(a^3) - x^2 sin(a^3)]e^{a^3}$$

Similarly,

$$\bar{\bar{x}}^2 = [x^1 sin(a^3) + x^2 cos(a^3)]e^{a^3}$$

Therefore, the given set of transformation satisfies the closure requirement.

2. The identity transformation is determined by the values $a^i{=}0$, i.e., for $a^i{=}0$

$$\bar{x}^1 = x^1 \qquad and \qquad \bar{x}^2 = x^2$$

3. Let

$$T_{a1}: \quad \bar{x}^1 = [x^1 cos(a_1^1) - x^2 sin(a_1^1)]e^{a_1^1}$$

$$\bar{x}^2 = [x^1 sin(a_1^1) + x^2 cos(a_1^1)]e^{a_1^1}$$

be a given transformation. The transformation defined by $a_1^1 = -a_1^1$ is the inverse transformation, since

$$T_{a1}^{-1} T_{a1}: \quad \bar{\bar{x}}^1 = [\bar{x}^1 cos(-a_1^1) - \bar{x}^2 sin(-a_1^1)]e^{-a_1^1}$$

$$\bar{\bar{x}}^2 = [\bar{x}^2 sin(-a_1^1) + \bar{x}^2 cos(-a_1^1)]e^{-a_1^1}$$

Therefore,

$$\bar{\bar{x}}^1 = [(x^1 cos(a_1^1) - x^2 sin(a_1^1))e^{a_1^1} cos(-a_1^1)$$

$$-(x^1 sin(a_1^1) + x^2 cos(a_1^1))e^{a_1^1} sin(-a_1^1)]e^{-a_1^1}$$

$$= [x^1(cos(a_1^1)cos(-a_1^1) - sin(a_1^1)sin(-a_1^1))$$

$$-x^2(sin(a_1^1)cos(-a_1^1) + cos(a_1^1)sin(-a_1^1))]e^{(a_1^1-a_1^1)}$$

$$= [x^1(cos(a_1^1 - a_1^1) - x^2 sin(a_1^1 - a_1^1)] = x^1$$

Similarly,

$$\bar{\bar{x}}^2 = [x^1 sin(a_1^1 - a_1^1) + x^2 cos(a_1^1 - a_1^1)] = x^2$$

Therefore, the inverse transformation

$$T_{a1}^{-1} T_{a1} x = Ix = x$$

is satisfied.

4. As a final step,we need to establish associativity. This can be shown quite readily as follows:

   $T_{a3}(T_{a2}T_{a1})x$:

$$\bar{\bar{\bar{x}}}^1 = (x^1 cos[a^3 + (a^2 + a^1)]$$
$$-x^2 sin[a^3 + (a^2 + a^1)])e^{(a^3+(a^2+a^1))}$$
$$\bar{\bar{\bar{x}}}^2 = (x^1 sin[a^3 + (a^2 + a^1)]$$
$$+x^2 cos[a^3 + (a^2 + a^1)])e^{(a^3+(a^2+a^1))}$$

Similarly,

$(T_{a3}T_{a2})T_{a1}x$:

$$\bar{\bar{\bar{x}}}^1 = (x^1 cos[(a^3 + a^2) + a^1] - x^2 sin[(a^3 + a^2)$$
$$+a^1])e^{[(a^3+a^2)+a^1]}$$
$$\bar{\bar{\bar{x}}}^2 = (x^1 sin[(a^3 + a^2) + a^1] + x^2 cos[(a^3 + a^2)$$
$$+a^1])e^{[(a^3+a^2)+a^1]}$$

Thus, the group property of a transformation has been established.

2.3 The Concept of an Infinitesimal Transformation

Having established the group character of a set of transformations, we now proceed to introduce the concept of an infinitesimal transformation. The discussion here will be limited to a one-parameter group.

A group is said to be continuous if its elements are identified by a set of continuous parameters. Thus, a set of transformation

$$\bar{x} = \phi(x, y, a) \quad ; \quad \bar{y} = \psi(x, y, a) \qquad (2.1)$$

is an example of a one-dimensional continuous group of transformation.

Since ϕ and ψ are continuous functions, the transformation can be written as

$$\bar{x} = \phi(x, y, a_0 + \epsilon) \quad ; \quad \bar{y} = \psi(x, y, a_0 + \epsilon) \qquad (2.2)$$

where a_0 is the value of the parameter corresponding to the identical transformation,i.e.,

$$\bar{x} = \phi(x, y, a_0) = x \quad ; \quad \bar{y} = \psi(x, y, a_0) = y \qquad (2.3)$$

and ϵ is an infinitesimal quantity which changes x and y by an infinitesimal amount and is defined as an infinitesimal transformation.

Expanding in Taylor series, Eq.(2.2) becomes

$$\bar{x} = \phi(x, y, a_0) + \frac{\epsilon}{1!}\Big(\frac{\partial \phi}{\partial a}\Big)_{a_0} + \frac{\epsilon^2}{2!}\Big(\frac{\partial^2 \phi}{\partial a^2}\Big)_{a_0} + \ldots\ldots\ldots\ldots \tag{2.4a}$$

and

$$\bar{y} = \psi(x, y, a_0) + \frac{\epsilon}{1!}\Big(\frac{\partial \psi}{\partial a}\Big)_{a_0} + \frac{\epsilon^2}{2!}\Big(\frac{\partial^2 \psi}{\partial a^2}\Big)_{a_0} + \ldots\ldots\ldots\ldots \tag{2.4b}$$

Since ϵ is an infinitesimal quantity, Eq.(2.4) then becomes

$$\bar{x} = x + \epsilon\xi(x, y) + O(\epsilon^2) \tag{2.5a}$$

$$\bar{y} = y + \epsilon\eta(x, y) + O(\epsilon^2) \tag{2.5b}$$

where

$$\xi(x, y) = \Big(\frac{\partial \phi}{\partial a}\Big)_{a_0}$$

$$\eta(x, y) = \Big(\frac{\partial \psi}{\partial a}\Big)_{a_0}$$

and the relation for the identical transformation, Eq.(2.3), has been used.

Eq.(2.1) is the "global transformation group" and Eq.(2.5) is the "infinitesimal transformation group".

2.4 Relation Between Global and Infinitesimal Groups of Transformations

A given function f(x,y) would be changed to $f(x_1, y_1)$ if it is subjected to the infinitesimal transformation defined by Eq.(2.5). Expanding in Taylor series, $f(x_1, y_1)$ becomes

$$f(x_1, y_1) = f(x + \epsilon\xi, y + \epsilon\eta)$$

$$= f(x, y) + \frac{\epsilon}{1!}Uf + \frac{\epsilon^2}{2!}U^2 f + \ldots\ldots\ldots \tag{2.6}$$

where

$$Uf = \xi\frac{\partial f}{\partial x} + \eta\frac{\partial f}{\partial y}$$

and $U^n f$ represents repeating the operator n times.

Example 2.1 Consider the infinitesimal transformation represented by

$$Uf = -y\frac{\partial f}{\partial x} + x\frac{\partial f}{\partial y}$$

This means that the transformation functions, Eq.(2.5), are given by $\xi(x,y) = -y; \eta(x,y) = x$.

In order to obtain the global group, the following subsystem must be solved:

$$\frac{d\bar{x}}{-\bar{y}} = \frac{d\bar{y}}{\bar{x}} = \frac{da}{1} \tag{2.7}$$

Eq.(2.7) can be rewritten as

$$\frac{d\bar{x}}{da} = -\bar{y} \quad ; \quad \frac{d\bar{y}}{da} = \bar{x} \tag{2.8}$$

The solutions of Eqs.(2.8) can be written as *

$$\bar{x} = c_1 cos(a) + c_2 sin(a) \tag{2.9a}$$

$$\bar{y} = c_3 cos(a) + c_4 sin(a) \tag{2.9b}$$

To determine the four constants, we first substitute Eqs. (2.9) into Eqs.(2.8) and get

$$c_4 = c_1 \quad and \quad -c_3 = c_2$$

The identity transformation is obtained by setting a=0. Eqs.(2.9) therefore give

$$c_1 = x \quad and \quad c_3 = y$$

The global group of transformation can be written as

$$\bar{x} = x\ cos(a) - y\ sin(a) \tag{2.10a}$$

$$\bar{y} = y\ cos(a) + x\ sin(a) \tag{2.10b}$$

Example 2.2 Consider the infinitesimal transformation represented by

$$Uf = c_1 x \frac{\partial f}{\partial x} + c_3 y \frac{\partial f}{\partial y}$$

The transformation functions are given by

$$\xi(x,y) = c_1 x \quad ; \quad \eta(x,y) = c_2 y$$

* Combining Eqs.(2.8), we obtain

$$\frac{d^2\bar{y}}{da^2} + \bar{y} = 0$$

To obtain the global group, the following subsystem must be solved:

$$\frac{d\bar{x}}{c_1\bar{x}} = \frac{d\bar{y}}{c_2\bar{y}} = \frac{da}{1} \tag{2.11}$$

Integrating Eq.(2.11), we get:

$$\bar{x} = \lambda_1 e^{c_1 a} \quad ; \quad \bar{y} = \lambda_2 e^{c_2 a} \tag{2.12}$$

To obtain the identity transformation, we set a = 0. Therefore, $\lambda_1 = x$ and $\lambda_2 = y$, and the global group of transformation is

$$\bar{x} = A^{c_1} x \quad ; \quad \bar{y} = A^{c_2} y \tag{2.13}$$

where $A = e^a$ is the parameter. Transformations represented by Eq.(2.13) form a "linear group of transformation".

Example 2.3 It can similarly shown that the infinitesimal operator

$$Uf = c_1 \frac{\partial f}{\partial x} + c_2 y \frac{\partial f}{\partial y}$$

would correspond to the one-parameter global "spiral group of transformation":

$$\bar{x} = x + c_1 A \quad ; \quad \bar{y} = y e^{c_2 A} \tag{2.14}$$

2.5 The Concept of Invariance

A function f(x,y) is said to be invariant under the infinitesimal transformation defined by Eq.(2.5), if it is unaltered by the transformation,i.e.,

$$f(\bar{x}, \bar{y}) = f(x, y) \tag{2.15}$$

Eq.(2.6) shows that Eq.(2.15) will be satisfied if Uf , U^2f , U^3f , etc. are simultaneously equal to zero. However, since $U^2 f$=U(Uf) and $U^3 f = U(U^2 f)$, it follows that the condition

$$Uf = \xi \frac{\partial f}{\partial x} + \eta \frac{\partial f}{\partial y} = 0 \tag{2.16}$$

is both necessary and sufficient requirement for invariance of f(x,y).

Based on the above theorem, the "invariant function" under a given group of transformations represented by Uf can be solved from Eq.(2.16). From elementary theories of partial differential equations [16], we have

$$\frac{dx}{\xi} = \frac{dy}{\eta} \tag{2.17}$$

The solution to this equation is given by $\Omega(x, y)$=constant, which is the required invariant function corresponding to an operator U. Since Eq.(2.17) has only one independent solution depending on a single arbitrary constant, a one-parameter group of transformations in two variables has one and only one independent invariant.

Example 2.4 Consider the rotation group given in Example 2.1. The solution to the equation

$$\frac{dx}{-y} = \frac{dy}{x}$$

is $x^2 + y^2$=constant, and is the invariant function of the rotation group.

Example 2.5 Consider the linear group given in Example 2.2. The solution of the equation

$$\frac{dx}{c_1 x} = \frac{dy}{c_2 y}$$

gives the invariant function

$$\frac{y}{x^{c_2/c_1}} = constant$$

Example 2.6 Consider the spiral group given in Example 2.3. The solution to the equation

$$\frac{dx}{c_1} = \frac{dy}{c_2 y}$$

gives the invariant function y/e^{cx} =constant where c=c_2/c_1.

The above concept can easily be generalized to n variables. For n variables, the condition for a function f(x_1,.....,x_n) to be invariant under a one-parameter group of transformation

$$\bar{x}_i = x_i + \epsilon\xi_i(x_1, ..., x_n) + O(\epsilon^2)$$

$$(i = 1, ..., n) \tag{2.18}$$

is

$$Uf = \xi_1(x_1, ..., x_n)\frac{\partial f}{\partial x_1} + ... + \xi_n(x_1, ..., x_n)\frac{\partial f}{\partial x_n} = 0 \tag{2.19}$$

The invariant functions can then be solved from the following system of equations

$$\frac{dx_1}{\xi_1} = \frac{dx_n}{\xi_n} \tag{2.20}$$

Since there exists (n-1) independent solutions to Eq.(2.20), it follows that a one-parameter group of transformations in n variables has (n-1) independent invariants.

2.6 Invariance of Differential Equations Under Groups of Transformations

A basic theorem on the determination of relationships satisfied by functions admitting a given group of infinitesimal transformation will now be given. Consider a function

$$\varphi = \varphi(x_1, ..., x_m; y_1, ..., y_n; ....; \frac{\partial^k y_1}{\partial {x_1}^k}, ..., \frac{\partial^k y_n}{\partial {x_n}^k})$$

the arguments of which, assumed p in number, contain derivatives of y_j up to order k. Such a function is known as a differential form of the $k-th$ order in m independent variables. If the arguments are designated by $z_1, ..., z_p$, where

$$z_1 = x_1$$

$$z_2 = x_2$$

$$..........$$

$$..........$$

$$z_{p-1} = \frac{\partial^k y_n}{\partial (x_{m-1})^k}$$

$$z_p = \frac{\partial^k y_n}{\partial (x_m)^k}$$

the function φ can be written in a simpler form as

$$\varphi = \varphi(z_1, ..., z_p) \tag{2.21}$$

The function φ is said to be invariant under the group of infinitesimal transformation defined by

$$\bar{z}_i = z_i + \epsilon \xi_i(z_1, ..., z_p) + O(\epsilon^2) \tag{2.22}$$

$$(i = 1, ..., p)$$

if $U\varphi = 0$, i.e. ,

$$\xi_1(z_1, ..., z_p)\frac{\partial \varphi}{\partial z_1} + ... + \xi_p(z_1, ..., z_p)\frac{\partial \varphi}{\partial z_p} = 0 \tag{2.23}$$

It was shown that for a group of transformations with p variables, Eq.(2.22),there are (p-1) functionally independent invariants:

$$\eta_m = \Omega_m(z_1, ..., z_p) = constant \tag{2.24}$$

$$(m = 1, ..., p-1)$$

satisfying Eq.(2.19),i.e.,

$$\xi_1 \frac{\partial \Omega_m}{\partial z_1} + ... + \xi_p \frac{\partial \Omega_m}{\partial z_p} = 0 \tag{2.25}$$

Now, if a change of variables is made of the given function φ, Eq.(2.21), from $(z_1, ..., z_p)$ to $(\eta_1, ..., \eta_{p-1}, z_p)$, we get

$$\varphi(z_1, ..., z_p) = \psi(\eta_1, ..., \eta_{p-1}; z_p) \tag{2.26}$$

The condition of invariance of φ, Eq.(2.23) becomes

$$\begin{aligned} \phi = \xi_1 (\frac{\partial \psi}{\partial \eta_1} \frac{\partial \eta_1}{\partial z_1} + ... + \frac{\partial \psi}{\partial z_p} \frac{\partial z_p}{\partial z_1}) + ... \\ + \xi_p (\frac{\partial \psi}{\partial \eta_1} \frac{\partial \eta_1}{\partial z_p} + ... + \frac{\partial \psi}{\partial z_p} \frac{\partial z_p}{\partial z_p}) \\ = \frac{\partial \psi}{\partial \eta_1} (\xi_1 \frac{\partial \eta_1}{\partial z_1} + ... + \xi_p \frac{\partial \eta_1}{\partial z_p}) + ... \\ + \frac{\partial \psi}{\partial \eta_{p-1}} (\xi_1 \frac{\partial \eta_{p-1}}{\partial z_1} + ... + \xi_p \frac{\partial \eta_{p-1}}{\partial z_p}) \\ + (\xi_1 + ... + \xi_p) \frac{\partial \psi}{\partial z_p} = 0 \end{aligned} \tag{2.27}$$

Based on Eqs.(2.24) and (2.25), all the terms in Eq.(2.27) except the last term are zero and, as a result, the following important conclusion is obtained:

$$\frac{\partial \psi(\eta_1, ..., \eta_{p-1}; z_p)}{\partial z_p} = 0$$

which means ψ is independent of z_p. Eq.(2.26) then becomes

$$\varphi(z_1, ..., z_p) = \psi(\eta_1, ..., \eta_{p-1}) \tag{2.28}$$

Thus, the conclusion that if φ is invariant, it is then expressed in terms of (p-1) functionally independent invariants,i.e.,

$$\varphi(z_1, ..., z_p) = \psi(\eta_1, ..., \eta_{p-1}) = 0 \tag{2.29}$$

As a result, the number of variables is reduced by one.

Morgan [10] proved the above theorem in a different manner. A brief outline will be given below. Assuming that y_j are the dependent variables and x_i the independent variables in the one-parameter group of transformations

$$T_a : \quad X_i = f_i(x_1, \ldots, x_m; a)$$
$$Y_j = h_j(y_1, \ldots, y_n; a),$$

the system of partial differential equations of order k

$$\varphi_j\left[x_1, \ldots, x_m; y_1, \ldots, y_n; \frac{\partial^k y_1}{\partial x_1{}^k}, \ldots, \frac{\partial^k y_n}{\partial x_m{}^k}\right] = 0$$

is invariant under this group of transformation, T_a,if each of the ϕ_j is "conformally invariant" under the transformation $T_a{}^k$ *. This means that

$$\varphi\left[x_1, \ldots, x_m; y_1, \ldots, y_n; \frac{\partial^k y_1}{\partial x_1{}^k}, \ldots, \frac{\partial^k y_n}{\partial x_m{}^k}\right]$$

$$= F\left[x_1, \ldots, x_m; y_1, \ldots, y_n; \ldots, \frac{\partial^k y_n}{\partial x_m{}^k}; a\right]$$

$$\cdot\varphi\left[x_1, \ldots, x_m; y_1, \ldots, y_n; \ldots, \frac{\partial^k y_n}{\partial (x_m)^k}\right] \tag{2.30}$$

If $F = f(a)$, φ_j is said to be "constant conformally invariant" under $T_a{}^k$. In particular, if F=f(a)=1, φ_j is said to be "absolutely invariant" under this group of transformation.

The above result leads to the theorem of *Morgan*[10]:

<u>*Theorem*</u>: Suppose that the forms φ_j are conformally invariant under the group $T_a{}^k$, then the invariant solutions of $\varphi_j = 0$ can be expressed in terms of the solutions of a new system of partial differential equations.

$$\varphi_j\left[\eta_1, \ldots, \eta_{m-1}; F_1, \ldots, F_n; \ldots, \frac{\partial^k F_n}{\partial (\eta_{m-1})^k}\right] = 0 \tag{2.31}$$

The η_i are the absolute invariants of the subgroup of transformations on the x_i alone and the variables F_j are such that

$$F_j(\eta_1, \ldots, \eta_{m-1}) = functions \;\; of \;\; (x_1, \ldots, x_m)$$

Although the same conclusions are obtained above, the method developed by Birkhoff [9] and Morgan [10] for a *given* group of transformations has been found to be the simplest to apply. The invocation of invariance has also led to (a) extremely successful deductive methods for obtaining various groups of transformations from a general form as evidenced by the series

* We append the transformation T_a by a set of transformations of the derivative of the y_j up to k. The enlarged group is called $T_a{}^k$.

of works by Moran and Gaggioli, and Na and Hansen, and Bluman and Cole presented in chapter 3, (b) the transformation of boundary value problems to initial value problems, covered in chapter 9, and (c) discovering mappings that transform nonlinear differential equations to linear differential equations as discussed in chapter 10.

2.7 The Extended Group of Transformations

For the one-parameter group of transformations

$$\bar{x} = \varphi(x, y, a) \quad ; \quad \bar{y} = \psi(x, y, a)$$

defined in Eq.(2.1), the differential coefficient p(=dy/dx) can be considered as a third variable which, under this group of transformations, will be transformed to $\bar{p}$ by the following transformation function:

$$\bar{p} = \frac{d\bar{y}}{d\bar{x}} = \frac{\frac{\partial \psi}{\partial x} + \frac{\partial \psi}{\partial y} \cdot p}{\frac{\partial \phi}{\partial x} + \frac{\partial \phi}{\partial y} \cdot p} = P(x, y, p; a) \tag{2.32}$$

It can be easily shown that the more general transformation

$$\bar{x} = \varphi(x, y, a) \quad ; \quad \bar{y} = \phi(x, y, a) \tag{2.33}$$

$$\bar{p} = P(x, y, p; a)$$

forms a group [17], which is known as the extended group of the group given in Eq.(2.1).

Next, the differential coefficient $\bar{p}$ for the infinitesimal transformation defined in Eq.(2.5) is sought. It can be shown that the expansions given in Eqs.(2.4) can be written as [17]

$$\bar{x} = x + \frac{\epsilon}{1!}\xi(x, y) + \frac{\epsilon^2}{2!}[\xi \frac{\partial \xi}{\partial x} + \eta \frac{\partial \xi}{\partial y}] + ... \tag{2.34a}$$

$$\bar{y} = y + \frac{\epsilon}{1!}\eta(x, y) + \frac{\epsilon^2}{2!}[\xi \frac{\partial \eta}{\partial x} + \eta \frac{\partial \eta}{\partial y}] + ... \tag{2.34b}$$

The differential coefficient $\bar{p}$ given in Eq.(2.32) then becomes

$$\bar{p} = \frac{dy + \frac{\epsilon}{1!}(\frac{\partial \eta}{\partial x} dx + \frac{\partial \eta}{\partial y} dy) + ...}{dx + \frac{\epsilon}{1!}(\frac{\partial \xi}{\partial x} dx + \frac{\partial \xi}{\partial y} dy) + ...}$$

$$= p + \frac{\epsilon}{1!}[\frac{\partial \eta}{\partial x} + (\frac{\partial \eta}{\partial y} - \frac{\partial \xi}{\partial x})p - \frac{\partial \xi}{\partial y} p^2] + ...$$

or,if higher-order terms of ϵ are omitted,

$$\bar{p} = p + \epsilon\xi(x, y, p)$$

where

$$\varsigma(x, y, p) = \frac{\partial \eta}{\partial x} + (\frac{\partial \eta}{\partial y} - \frac{\partial \xi}{\partial x})p - \frac{\partial \xi}{\partial y}p^2 \tag{2.35}$$

Thus, the group of transformations defined by Eq.(2.5) and Eq.(2.35) simultaneously,i.e.,

$$\begin{aligned} \bar{x} &= x + \epsilon\xi(x, y) + O(\epsilon^2) \\ \bar{y} &= y + \epsilon\eta(x, y) + O(\epsilon^2) \\ \bar{p} &= p + \epsilon\varsigma(x, y, p) + O(\epsilon^2) \end{aligned} \tag{2.36}$$

form a group known as the "extended infinitesimal group of transformations". Note here that p is considered as an independent variable.

Extension of the concept to higher order derivatives can be made based on the same reasoning. Consider now the second- order derivative,

$$q = \frac{dp}{dx} = \frac{d^2y}{dx^2},$$

which may be considered as an independent variable. Under the group of transformation defined by Eq.(2.33), q will be transformed to $\bar{q}$ by the following relationship:

$$\begin{aligned} \bar{q} = \frac{d\bar{p}}{d\bar{x}} &= \frac{\frac{\partial P}{\partial x} + \frac{\partial P}{\partial y}p + \frac{\partial P}{\partial p}q}{\frac{\partial \phi}{\partial x} + p\frac{\partial \phi}{\partial y}} \\ &= Q(x, y, p, q; a) \end{aligned} \tag{2.37}$$

The group of transformation defined by

$$\bar{x} = \phi(x, y, a) \quad ; \quad \bar{y} = \psi(x, y, a);$$

$$\bar{p} = P(x, y, p, a) \quad ; \quad \bar{q} = Q(x, y, p, q, a) \tag{2.38}$$

is known as the "twice-extended group of transformations".

We can write the twice extended group of infinitesimal transformations as follows:

$$\begin{aligned} \bar{x} &= x + \epsilon\xi(x, y) + O(\epsilon^2) \\ \bar{y} &= y + \epsilon\eta(x, y) + O(\epsilon^2) \\ \bar{p} &= p + \epsilon\varsigma(x, y, p) + O(\epsilon^2) \\ \bar{q} &= q + \epsilon\delta(x, y, p, q) + O(\epsilon^2) \end{aligned} \tag{2.39}$$

where

$$\delta(x,y,p,q) = \frac{\partial\varsigma}{\partial x} + p\frac{\partial\varsigma}{\partial y} + q\frac{\partial\varsigma}{\partial p} - q(\frac{\partial\xi}{\partial x} + p\frac{\partial\xi}{\partial y}) \qquad (2.40)$$

The concept can be extended to higher-order derivatives. Three examples on the concept of extended group are given below.

Example 2.7 For the rotation group represented by

$$Uf = -y\frac{\partial f}{\partial x} + x\frac{\partial f}{\partial y}$$

it can be easily shown that the extended group of infinitesimal transformations is

$$\bar{x} = x + \epsilon\xi(x,y) + O(\epsilon^2)$$

$$\bar{y} = y + \epsilon\eta(x,y) + O(\epsilon^2)$$

$$\bar{p} = p + \epsilon\varsigma(x,y,p) + O(\epsilon^2)$$

where $\xi = -y, \eta = x$, and from Eq.(2.35),

$$\varsigma = \frac{\partial\eta}{\partial x} + (\frac{\partial\eta}{\partial y} - \frac{\partial\xi}{\partial x})p + \frac{\partial\xi}{\partial y}p^2 = 1 + p^2$$

The symbol for this extended group of transformation is therefore

$$Uf = -y\frac{\partial f}{\partial x} + x\frac{\partial f}{\partial y} + (1+p^2)\frac{\partial f}{\partial p}$$

Example 2.8 For the linear group , the extended group of transformation can be shown to be represented by

$$Uf = c_1 x\frac{\partial f}{\partial x} + c_2 y\frac{\partial f}{\partial y} + (c_2 - c_1)p\frac{\partial f}{\partial p}$$

Example 2.9 For the spiral group, the extended group of transformations can be shown to be represented by

$$Uf = c_1\frac{\partial f}{\partial x} + c_2\frac{\partial f}{\partial y} + c_2 p\frac{\partial f}{\partial p}$$

2.8 The Characteristic Function

In this section, we will express the functions

$$\xi(x,y) \ ; \ \eta(x,y) \ ; \ \varsigma(x,y,p) \ ; \ \delta(x,y,p,q)$$

in terms of a single function. Let us define an infinitesimal group of transformations

$$\bar{x} = x + \epsilon\xi(x, y) + O(\epsilon^2)$$

$$\bar{y} = y + \epsilon\eta(x, y) + O(\epsilon^2) \tag{2.5}$$

where ξ and η are the transformation functions of the group.

Let $y = y(x)$ be a solution of the ordinary differential equation

$$My = 0 \tag{2.41}$$

where M is a linear or nonlinear differential operator. If the differential equation is invariant under Eq.(2.41) the solution must map into itself,i.e.,

$$y(\bar{x}) = \bar{y}(x, y, \epsilon) \tag{2.42}$$

In terms of the infinitesimal transformation, Eq.(2.42) can be written as:

$$y(x + \epsilon\xi) = y(x) + \epsilon\eta(x, y) + O(\epsilon^2) \tag{2.43}$$

Expanding the left side of Eq.(2.43) and equating coefficients of ϵ, we get:

$$\xi(x, y)\frac{dy}{dx} - \eta(x, y) = 0 \tag{2.44}$$

Eq.(2.44) is called the "invariant surface equation".

Solution of Eq.(2.44) would lead to the following transformation:

$$y(x) = F(x; \eta, f(\eta))$$

In order to express ξ and η in terms of a single function called the "characteristic function",W, we rewrite Eq.(2.44) as:

$$W = \xi p - \eta \tag{2.45}$$

where p=dy/dx. It is seen that the characteristic function is indeed the invariant surface.

From Eq.(2.45), it can be seen that

$$\xi = \frac{\partial W}{\partial p} \quad ; \quad \eta = p\frac{\partial W}{\partial p} - W \tag{2.46}$$

For the extended group of transformation,

$$\bar{p} = p + \epsilon\varsigma(x, y, p) + O(\epsilon^2)$$

$$\bar{q} = q + \epsilon\delta(x, y, p, q) + O(\epsilon^2) \tag{2.47}$$

For the function ς, Eq.(2.35) gives:

$$\varsigma(x,y,p) = p\frac{\partial^2 W}{\partial x\partial p} - \frac{\partial W}{\partial x} + [p\frac{\partial^2 W}{\partial y\partial p} - \frac{\partial W}{\partial y} - \frac{\partial^2 W}{\partial x\partial p}] - \frac{\partial^2 W}{\partial y\partial p}p^2 \tag{2.48}$$

or,

$$\varsigma(x,y,p) = -\frac{\partial W}{\partial x} - p\frac{\partial W}{\partial y} = -X(W) \tag{2.49}$$

where the operator

$$X(\;) = \frac{\partial(\;)}{\partial x} + p\frac{\partial(\;)}{\partial y}$$

For the function δ, Eq.(2.40) gives

$$\begin{aligned}\delta(x,y,p,q) &= -(\frac{\partial^2 W}{\partial x^2} + p\frac{\partial^2 W}{\partial x\partial y}) - p(\frac{\partial^2 W}{\partial x\partial y} + p\frac{\partial^2 W}{\partial y^2}) \\ &\quad -q(\frac{\partial W}{\partial y} + \frac{\partial^2 W}{\partial p\partial x} + p\frac{\partial^2 W}{\partial p\partial y}) - q(\frac{\partial^2 W}{\partial x\partial p} + p\frac{\partial^2 W}{\partial y\partial p}) \\ &= -(\frac{\partial^2 W}{\partial x^2} + 2p\frac{\partial^2 W}{\partial x\partial y} + p^2\frac{\partial^2 W}{\partial y^2}) - q\frac{\partial W}{\partial y} \\ &\quad -2q(\frac{\partial^2 W}{\partial x\partial p} + p\frac{\partial^2 W}{\partial y\partial p})\end{aligned} \tag{2.50}$$

or,

$$\delta(x,y,p,q) = -(X^2 + 2qX\frac{\partial}{\partial p} + q\frac{\partial}{\partial y})\,W \tag{2.51}$$

where

$$X^2 = \frac{\partial^2(\;)}{\partial x^2} + 2p\frac{\partial^2(\;)}{\partial x\partial y} + p^2\frac{\partial^2(\;)}{\partial y^2}$$

The same approach can be followed to obtain the transformation function * of the third-order derivative, r$(=d^3y/dx^3)$, where

$$\bar{r} = r + \epsilon\rho(x,y,p,q,r) + O(\epsilon^2) \tag{2.52}$$

with

$$\begin{aligned}\rho(x,y,p,q,r) &= \frac{\partial\delta}{\partial x} + \frac{\partial\delta}{\partial y}p \\ &\quad + \frac{\partial\delta}{\partial p}q + \frac{\partial\delta}{\partial q}r - r(\frac{\partial\xi}{\partial x} + \frac{\partial\xi}{\partial y}p)\end{aligned} \tag{2.53}$$

* In this book, the terms "transformation functions" and "infinitesimals of a group" are used interchangeably.

In terms of the characteristic function, W, we have

$$-\rho = X^3 W + 3qX^2(\frac{\partial W}{\partial p}) + 3qX(\frac{\partial W}{\partial y}) + 3q^2(\frac{\partial^2 W}{\partial y \partial p})$$

$$+r[3X(\frac{\partial W}{\partial p}) + \frac{\partial W}{\partial y}] \tag{2.54}$$

where $X(\ )$ and $X^2(\ )$ are defined in Eqs.(2.49) and (2.51), and

$$X^3(\ ) = \frac{\partial^3(\ )}{\partial x^3} + 2p\frac{\partial^3(\ )}{\partial x^2 \partial y} + p^2\frac{\partial^3(\ )}{\partial x \partial y^2}$$

$$+p[\frac{\partial^3(\ )}{\partial y \partial x^2} + 2p\frac{\partial^3(\ )}{\partial x \partial y^2} + p^2\frac{\partial^3(\ )}{\partial y^3}] \tag{2.55}$$

2.9 Transformations Involving Two Independent Variables

We will now consider transformation in which the number of independent variables is two. Such transformations are needed if partial differential equations are to be treated. Choosing t and x as the independent variables and u as the dependent variable, we now introduce the infinitesimal transformation

$$\begin{aligned} \bar{t} &= t + \epsilon\alpha(t, x, u) + O(\epsilon^2) \\ \bar{x} &= x + \epsilon\beta(t, x, u) + O(\epsilon^2) \\ \bar{u} &= u + \epsilon\varsigma(t, x, u) + O(\epsilon^2), \end{aligned} \tag{2.56}$$

where α,β and ς are the transformation functions. Let $u = u(x, t)$ be a solution of the partial differential equation

$$Mu = 0 \tag{2.57}$$

where M is a nonlinear or linear differential operator. If the differential equation, Eq.(2.57), is invariant under the group of transformation, Eq.(2.56), the solution must map into itself, i.e.,

$$u(\bar{x}, \bar{t}) = \bar{u}(x, t, u, \epsilon) \tag{2.58}$$

In terms of the transformation functions, Eq.(2.58) can be written as:

$$u(x + \epsilon\beta, t + \epsilon\alpha) = u(x, t) + \epsilon\varsigma(x, t, u) + O(\epsilon^2) \tag{2.59}$$

Expanding the left-hand side of Eq.(2.59) and equating the coefficients of ϵ, we get:

$$\beta(u, x, t)\frac{\partial u}{\partial x} + \alpha(u, x, t)\frac{\partial u}{\partial t} = \varsigma(u, x, t) \tag{2.60}$$

Eq.(2.60) is the "invariant surface condition". The solution of Eq.(2.60) would lead to the similarity transformation

$$\eta(x,t) = constant$$

$$u(x,t) = F(x,t,\eta,f(\eta)).$$

In order to express α,β and ς in terms of the characteristic function, W,we rewrite Eq.(2.60) as:

$$W = \alpha p + \beta q - \varsigma \tag{2.61}$$

where

$$p = \frac{\partial u}{\partial t} \quad ; \quad q = \frac{\partial u}{\partial x}$$

It can be seen that the characteristic function is the invariant surface equation. Using Eq.(2.61), we have

$$\alpha = \frac{\partial W}{\partial p}$$

$$\beta = \frac{\partial W}{\partial q} \tag{2.62}$$

$$\varsigma = p\frac{\partial W}{\partial p} + q\frac{\partial W}{\partial q} - W$$

Eqs.(2.62) give α,β and ς as functions of the characteristic function, W.

Next, the group is extended to include the differential coefficients p and q, i.e.,

$$\bar{p} = p + \epsilon\pi_1(t,x,u,p,q) + O(\epsilon^2)$$

$$\bar{q} = q + \epsilon\pi_2(t,x,u,p,q) + O(\epsilon^2) \tag{2.63}$$

The extended transformation functions or infinitesimals of the group, π_1 and π_2 can be expressed in terms of α, β and ς. Details are as follows:

$$\frac{\partial t}{\partial \bar{t}} = \frac{\partial}{\partial \bar{t}}[\bar{t} - \epsilon\alpha(t,x,u)] + O(\epsilon^2)$$

$$= 1 - \epsilon[\frac{\partial \alpha}{\partial t} + \frac{\partial \alpha}{\partial u}p]\frac{\partial t}{\partial \bar{t}} + O(\epsilon^2)$$

from which,

$$\frac{\partial t}{\partial \bar{t}} = \frac{1}{1 + \epsilon(\frac{\partial \alpha}{\partial t} + \frac{\partial \alpha}{\partial u}p)} + O(\epsilon^2)$$

$$= 1 - \epsilon(\frac{\partial \alpha}{\partial t} + \frac{\partial \alpha}{\partial u}p) + O(\epsilon^2) \tag{2.64}$$

Similarly, we have

$$\frac{\partial x}{\partial \bar{x}} = \frac{\partial}{\partial \bar{x}}[\bar{x} - \epsilon\beta(t, x, u)] + O(\epsilon^2)$$

$$= 1 - \epsilon(\frac{\partial \beta}{\partial x} + \frac{\partial \beta}{\partial u}q)\frac{\partial x}{\partial \bar{x}} + O(\epsilon^2)$$

from which

$$\frac{\partial x}{\partial \bar{x}} = 1 - \epsilon(\frac{\partial \beta}{\partial x} + \frac{\partial \beta}{\partial u}q) + O(\epsilon^2) \tag{2.65}$$

Next, we have

$$\frac{\partial t}{\partial \bar{x}} = \frac{\partial}{\partial \bar{x}}[\bar{t} - \epsilon\alpha(t, x, u) + O(\epsilon^2)]$$

$$= -\epsilon\frac{\partial}{\partial \bar{x}}[\alpha(t, x, u)] + O(\epsilon^2)$$

$$= -\epsilon(\frac{\partial \alpha}{\partial x} + \frac{\partial \alpha}{\partial u}q)\frac{\partial x}{\partial \bar{x}} + O(\epsilon^2)$$

Therefore,

$$\frac{\partial t}{\partial \bar{x}} = -\epsilon(\frac{\partial \alpha}{\partial x} + \frac{\partial \alpha}{\partial u}q) + O(\epsilon^2) \tag{2.66}$$

Finally, we get

$$\frac{\partial x}{\partial \bar{t}} = \frac{\partial}{\partial \bar{t}}[\bar{x} - \epsilon\beta(t, x, u) + O(\epsilon^2)]$$

$$= -\epsilon\frac{\partial}{\partial \bar{t}}[\beta(t, x, u)] + O(\epsilon^2)$$

$$= -\epsilon(\frac{\partial \beta}{\partial t} + \frac{\partial \beta}{\partial u}p) + O(\epsilon^2) \tag{2.67}$$

Using Eqs.(2.64) to (2.67) and the chain rule of differentiation, we can obtain the derivatives:

$$\frac{\partial \bar{u}}{\partial \bar{t}} = \frac{\partial}{\partial \bar{t}}[u + \epsilon\varsigma(t, x, u) + O(\epsilon^2)]$$

$$= \frac{\partial}{\partial t}[u + \epsilon\varsigma(t, x, u)]\frac{\partial t}{\partial \bar{t}} + \frac{\partial}{\partial x}[u + \epsilon\varsigma(t, x, u)]\frac{\partial x}{\partial \bar{t}} + O(\epsilon^2)$$

$$= [p + \epsilon(\frac{\partial \varsigma}{\partial t} + \frac{\partial \varsigma}{\partial u}p)][1 - \epsilon(\frac{\partial \alpha}{\partial t} + \frac{\partial \alpha}{\partial u}p)]$$

$$+[q + \epsilon(\frac{\partial \varsigma}{\partial x} + \frac{\partial \varsigma}{\partial u}q)][-\epsilon(\frac{\partial \beta}{\partial t} + \frac{\partial \beta}{\partial u}p)] + O(\epsilon^2)$$

Therefore,

$$\bar{p} = \frac{\partial \bar{u}}{\partial \bar{t}} = p + \epsilon[\frac{\partial \varsigma}{\partial t} + \frac{\partial \varsigma}{\partial u}p - p(\frac{\partial \alpha}{\partial t} + \frac{\partial \alpha}{\partial u}p)$$

$$-q(\frac{\partial\beta}{\partial t}+\frac{\partial\beta}{\partial u}p)]+O(\epsilon^2) \tag{2.68}$$

and

$$\frac{\partial\bar{u}}{\partial\bar{x}}=\frac{\partial}{\partial\bar{x}}[u+\epsilon\varsigma(t,x,u)+O(\epsilon^2)]$$

$$=\frac{\partial}{\partial t}[u+\epsilon\varsigma(t,x,u)]\frac{\partial t}{\partial\bar{x}}+\frac{\partial}{\partial x}[u+\epsilon\varsigma(t,x,u)\frac{\partial x}{\partial\bar{x}}+O(\epsilon^2)]$$

$$=[p+\epsilon(\frac{\partial\varsigma}{\partial t}+\frac{\partial\varsigma}{\partial u}q)][-\epsilon(\frac{\partial\alpha}{\partial x}+\frac{\partial\alpha}{\partial u}q)]$$

$$+[q+\epsilon(\frac{\partial\varsigma}{\partial x}+\frac{\partial\varsigma}{\partial u}q)][1-\epsilon(\frac{\partial\beta}{\partial x}+\frac{\partial\beta}{\partial u}q)]+O(\epsilon^2)$$

Simplifying,

$$\bar{q}=\frac{\partial\bar{u}}{\partial\bar{x}}=q+\epsilon[\frac{\partial\varsigma}{\partial x}+\frac{\partial\varsigma}{\partial u}q-q(\frac{\partial\beta}{\partial x}+\frac{\partial\beta}{\partial u}q)]+O(\epsilon^2) \tag{2.69}$$

The extended transformation functions π_1 and π_2 can now be written as

$$\pi_1=\frac{\partial\varsigma}{\partial t}+(\frac{\partial\varsigma}{\partial u}-\frac{\partial\alpha}{\partial t})p-\frac{\partial\alpha}{\partial u}p^2$$

$$-q(\frac{\partial\beta}{\partial t}+\frac{\partial\beta}{\partial u}p) \tag{2.70}$$

$$\pi_2=\frac{\partial\varsigma}{\partial x}+(\frac{\partial\varsigma}{\partial u}-\frac{\partial\beta}{\partial x})q-\frac{\partial\beta}{\partial u}q^2$$

$$-p(\frac{\partial\alpha}{\partial x}+\frac{\partial\alpha}{\partial u}q) \tag{2.71}$$

In terms of the characteristic function, W, the above equations become:

$$\pi_1=-\frac{\partial W}{\partial t}-p\frac{\partial W}{\partial u} \tag{2.72}$$

$$\pi_2=-\frac{\partial W}{\partial x}-q\frac{\partial W}{\partial u} \tag{2.73}$$

Higher-order derivatives can be derived in a similar manner.

2.10 Transformation Involving Two Dependent and Two Independent Variables

We will now consider transformation in which the number of dependent and independent variables are two. Let the dependent variables be u(t,x) and v(t,x). We introduce the infinitesimal transformation

$$\bar{t}=t+\epsilon\alpha(t,x,u,v)+O(\epsilon^2)$$

$$\bar{x} = x + \epsilon\beta(t, x, u, v) + O(\epsilon^2)$$
$$\bar{u} = u + \epsilon\varsigma(t, x, u, v) + O(\epsilon^2) \qquad (2.74)$$
$$\bar{v} = v + \epsilon\delta(t, x, u, v) + O(\epsilon^2)$$

Let u(x,t) and v(x,t) be the solutions of partial differential equations:

$$M_i(u, v) = 0 \quad ; \quad i = 1, 2 \qquad (2.75)$$

where M_i can be linear or nonlinear differential operator. If the differential equations, Eq.(2.75), are invariant under the group of transformations defined by Eq.(2.74), the solutions must map into themselves,i.e.,

$$u(\bar{x}, \bar{t}) = \bar{u}(x, t, u, v; \epsilon)$$
$$v(\bar{x}, \bar{t}) = \bar{v}(x, t, u, v; \epsilon) \qquad (2.76)$$

In terms of the infinitesimals, Eq.(2.76) can be written as:

$$u(x + \epsilon\beta; t + \epsilon\alpha) = u(x, t) + \epsilon\varsigma(t, x, u, v) + O(\epsilon^2) \qquad (2.77a)$$

$$v(x + \epsilon\beta; t + \epsilon\alpha) = v(x, t) + \epsilon\delta(t, x, u, v) + O(\epsilon^2) \qquad (2.77b)$$

Expanding the left-hand sides of Eqs.(2.77) and equating the coefficients of ϵ, we get:

$$\beta(x, t, u, v)\frac{\partial u}{\partial x} + \alpha(x, t, u, v)\frac{\partial u}{\partial t} - \varsigma(t, x, u, v) = 0 \qquad (2.78a)$$

$$\beta(x, t, u, v)\frac{\partial v}{\partial x} + \alpha(x, t, u, v)\frac{\partial v}{\partial t} - \delta(t, x, u, v) = 0 \qquad (2.78b)$$

Eqs.(2.78) are the invariant surface conditions, and lead to the similarity transformation

$$\eta(x, t) = constant$$
$$u(x, t) = F(x, t, \eta, f(\eta))$$
$$v(x, t) = G(x, t, \eta, g(\eta))$$

In order to express α, β, ς and δ in terms of two characteristic functions, W_1 and W_2, we rewrite Eq.(2.78) as

$$W_1 = \beta p_{12} + \alpha p_{11} - \varsigma$$

$$W_2 = \beta p_{22} + \alpha p_{21} - \delta \qquad (2.79)$$

where

$$p_{11} = \frac{\partial u}{\partial t} \quad ; \quad p_{12} = \frac{\partial u}{\partial x}$$

$$p_{21} = \frac{\partial v}{\partial t} \quad ; \quad p_{22} = \frac{\partial v}{\partial x}.$$

It can again be seen that the characteristic functions are the invariant surface equation, rearranged.

From Eqs.(2.7), it can be seen that

$$\alpha = \frac{\partial W_1}{\partial p_{11}} = \frac{\partial W_2}{\partial p_{21}} \quad ; \quad \beta = \frac{\partial W_1}{\partial p_{12}} = \frac{\partial W_2}{\partial p_{22}}$$

$$\varsigma = p_{11}\frac{\partial W_1}{\partial p_{11}} + p_{12}\frac{\partial W_1}{\partial p_{12}} - W_1 \tag{2.80}$$

$$\delta = p_{21}\frac{\partial W_2}{\partial p_{21}} + p_{22}\frac{\partial W_2}{\partial p_{22}} - W_2$$

Eqs.(2.80) express the transformation functions or the infinitesimals of the group, α, β, ς and δ in terms of the characteristic functions W_1 and W_2.

Next, the group is extended to include the coefficients p_{11},p_{12},p_{21} and p_{22}, namely,

$$\bar{p}_{11} = p_{11} + \pi_{11}(t, x, u, v, p_{11}, p_{12}, p_{21}, p_{22}) + O(\epsilon^2)$$

$$\bar{p}_{12} = p_{12} + \pi_{12}(t, x, u, v, p_{11}, p_{12}, p_{21}, p_{22}) + O(\epsilon^2)$$

$$\bar{p}_{21} = p_{21} + \pi_{21}(t, x, u, v, p_{11}, p_{12}, p_{21}, p_{22}) + O(\epsilon^2)$$

$$\bar{p}_{22} = p_{22} + \pi_{22}(t, x, u, v, p_{11}, p_{12}, p_{21}, p_{22}) + O(\epsilon^2)$$

(2.81*a, b, c, d*)

The extended transformation functions π_{11},π_{12},π_{21} and π_{22} can be expressed in terms of α,β,ς and δ. Details are given below:

$$\frac{\partial t}{\partial \bar{t}} = \frac{\partial}{\partial \bar{t}}[\bar{t} - \epsilon\alpha(t, x, u, v) + O(\epsilon^2)]$$

$$= 1 - \epsilon[\frac{\partial \alpha}{\partial t} + \frac{\partial \alpha}{\partial u}p_{11} + \frac{\partial \alpha}{\partial v}p_{21}]\frac{\partial t}{\partial \bar{t}} + O(\epsilon^2)$$

from which,

$$\frac{\partial t}{\partial \bar{t}} = \frac{1}{1 + \epsilon[\frac{\partial \alpha}{\partial t} + \frac{\partial \alpha}{\partial u}p_{11} + \frac{\partial \alpha}{\partial v}p_{21}]} + O(\epsilon^2)$$

or,

$$\frac{\partial t}{\partial \bar{t}} = 1 - \epsilon[\frac{\partial \alpha}{\partial t} + \frac{\partial \alpha}{\partial u}p_{11} + \frac{\partial \alpha}{\partial v}p_{21}] + O(\epsilon^2) \tag{2.82}$$

Similarly, we get

$$\frac{\partial x}{\partial \bar{x}} = 1 - \epsilon[\frac{\partial \beta}{\partial x} + \frac{\partial \beta}{\partial u}p_{12} + \frac{\partial \beta}{\partial v}p_{22}] + O(\epsilon^2) \tag{2.83}$$

Next, we have

$$\frac{\partial t}{\partial \bar{x}} = \frac{\partial}{\partial \bar{x}}[\bar{t} - \epsilon\alpha(t, x, u, v)] + O(\epsilon^2)$$

$$= -\epsilon[\frac{\partial \alpha}{\partial x} + \frac{\partial \alpha}{\partial u}p_{12} + \frac{\partial \alpha}{\partial v}p_{22}]\frac{\partial x}{\partial \bar{x}} + O(\epsilon^2)$$

$$= -\epsilon[\frac{\partial \alpha}{\partial x} + \frac{\partial \alpha}{\partial u}p_{12} + \frac{\partial \alpha}{\partial v}p_{22}] + O(\epsilon^2) \tag{2.84}$$

and similarly

$$\frac{\partial x}{\partial \bar{t}} = -\epsilon[\frac{\partial \beta}{\partial t} + \frac{\partial \beta}{\partial u}p_{11} + \frac{\partial \beta}{\partial v}p_{21}] + O(\epsilon^2) \tag{2.85}$$

With the above relations, we can now obtain the transformed derivatives as follows:

$$\frac{\partial \bar{u}}{\partial \bar{t}} = \frac{\partial}{\partial \bar{t}}[u + \epsilon\varsigma(t, x, u, v) + O(\epsilon^2)]$$

$$= \frac{\partial}{\partial t}[u + \epsilon\varsigma]\frac{\partial t}{\partial \bar{t}} + \frac{\partial}{\partial x}[u + \epsilon\varsigma]\frac{\partial x}{\partial \bar{t}} + O(\epsilon^2)$$

$$= [p_{11} + \epsilon(\frac{\partial \varsigma}{\partial t} + \frac{\partial \varsigma}{\partial u}p_{11} + \frac{\partial \varsigma}{\partial v}p_{21})]\cdot$$

$$[1 - \epsilon(\frac{\partial \alpha}{\partial t} + \frac{\partial \alpha}{\partial u}p_{11} + \frac{\partial \alpha}{\partial v}p_{21})]$$

$$+[p_{12} + \epsilon(\frac{\partial \varsigma}{\partial x} + \frac{\partial \varsigma}{\partial u}p_{12} + \frac{\partial \varsigma}{\partial v}p_{22})]\cdot$$

$$[-\epsilon(\frac{\partial \beta}{\partial t} + \frac{\partial \beta}{\partial u}p_{11} + \frac{\partial \beta}{\partial v}p_{21})] + O(\epsilon^2)$$

Therefore,

$$\bar{p}_{11} = \frac{\partial \bar{u}}{\partial \bar{t}} = p_{11} + \epsilon[(\frac{\partial \varsigma}{\partial t} + \frac{\partial \varsigma}{\partial u}p_{11} + \frac{\partial \varsigma}{\partial v}p_{21})$$

$$-p_{11}(\frac{\partial \alpha}{\partial t} + \frac{\partial \alpha}{\partial u}p_{11} + \frac{\partial \alpha}{\partial v}p_{21})$$

$$-p_{12}(\frac{\partial \beta}{\partial t} + \frac{\partial \beta}{\partial u}p_{12} + \frac{\partial \beta}{\partial v}p_{22})] + O(\epsilon^2) \tag{2.86}$$

Similarly,

$$\bar{p}_{12} = \frac{\partial \bar{u}}{\partial \bar{x}} = p_{12} + \epsilon[(\frac{\partial \varsigma}{\partial x} + \frac{\partial \varsigma}{\partial u}p_{12} + \frac{\partial \varsigma}{\partial v}p_{22})$$

$$-p_{11}(\frac{\partial \alpha}{\partial x} + \frac{\partial \alpha}{\partial u}p_{12} + \frac{\partial \alpha}{\partial v}p_{22})$$

$$-p_{12}(\frac{\partial \beta}{\partial x} + \frac{\partial \beta}{\partial u}p_{12} + \frac{\partial \beta}{\partial v}p_{22})] + O(\epsilon^2) \tag{2.87}$$

Now,

$$\frac{\partial \bar{v}}{\partial \bar{t}} = \frac{\partial}{\partial t}[v + \epsilon\delta(t, x, u, v) + O(\epsilon^2)]$$

$$= \frac{\partial}{\partial t}(v + \epsilon\delta)\frac{\partial t}{\partial \bar{t}} + \frac{\partial}{\partial x}(v + \epsilon\delta)\frac{\partial x}{\partial \bar{x}} + O(\epsilon^2)$$

$$= [p_{21} + \epsilon(\frac{\partial \delta}{\partial t} + \frac{\partial \delta}{\partial u}p_{11} + \frac{\partial \delta}{\partial v}p_{21})]\cdot$$

$$[1 - \epsilon(\frac{\partial \alpha}{\partial t} + \frac{\partial \alpha}{\partial u}p_{11} + \frac{\partial \alpha}{\partial v}p_{21})]$$

$$+[p_{22} + \epsilon(\frac{\partial \delta}{\partial t} + \frac{\partial \delta}{\partial u}p_{12} + \frac{\partial \delta}{\partial v}p_{22})]\cdot$$

$$[-\epsilon(\frac{\partial \beta}{\partial t} + \frac{\partial \beta}{\partial u}p_{11} + \frac{\partial \beta}{\partial v}p_{21})] + O(\epsilon^2)$$

Therefore,

$$\bar{p}_{21} = \frac{\partial \bar{v}}{\partial \bar{t}} = p_{21} + \epsilon[(\frac{\partial \delta}{\partial t} + \frac{\partial \delta}{\partial u}p_{11} + \frac{\partial \delta}{\partial v}p_{21})$$

$$-p_{21}(\frac{\partial \alpha}{\partial t} + \frac{\partial \alpha}{\partial u}p_{11} + \frac{\partial \alpha}{\partial v}p_{21})$$

$$-p_{22}(\frac{\partial \beta}{\partial t} + \frac{\partial \beta}{\partial u}p_{11} + \frac{\partial \beta}{\partial v}p_{21})] + O(\epsilon^2) \tag{2.88}$$

In a similar manner,

$$\bar{p}_{22} = \frac{\partial \bar{v}}{\partial \bar{x}} = p_{22} + \epsilon[(\frac{\partial \delta}{\partial x} + \frac{\partial \delta}{\partial u}p_{12} + \frac{\partial \delta}{\partial v}p_{22})$$

$$-p_{21}(\frac{\partial \alpha}{\partial x} + \frac{\partial \alpha}{\partial u}p_{12} + \frac{\partial \alpha}{\partial v}p_{22})$$

$$-p_{22}(\frac{\partial \beta}{\partial x} + \frac{\partial \beta}{\partial u}p_{12} + \frac{\partial \beta}{\partial v}p_{22})] + O(\epsilon^2) \tag{2.89}$$

Comparing Eqs.(2.81) with Eqs.(2.86) to (2.89), and equating the coefficients of ϵ, we get the extended transformation functions or the infinitesimals of the group, π_{11},π_{12},π_{21} and π_{22}. These transformation functions can be expressed in terms of two characteristic functions, W_1 and W_2, by substituting Eqs.(2.80) into expressions for π_{11}, π_{12}, π_{21} and π_{22}. Due to their complexity, we will not repeat them here.

2.11 Dimensional and Affine Groups of Transformations

The dimensional group of transformations Γ, can be subsumed in the general group of transformations defined by Eq.(2.1). The dimensional group which can be expressed as

$$\Gamma : \qquad \bar{Z}_i = a^{\gamma_{i1}} \cdots a^{\gamma_{ir}} Z_i \tag{2.90}$$

$$(i = 1, \ldots\ldots, m)$$

is associated with the concept of "dimensions". For $\Gamma \leq m$ the parameters of the group Γ are non-essential if and only if the rank σ of the matrix of exponents $|\Gamma_{i\alpha}|$ (i=1,....,m;α=1,....,r) is less than r;$\sigma < r$. For $m < r$, however,the parameters are always non-essential.

For each i, the exponents $\gamma_{i\alpha}$ (α=1,....,r) of Γ are termed "dimensions". If $\Gamma_{i\alpha}$ =0 for any i, and all α, then the physical concept associated with z_i is said to be dimensionless with respect to Γ. The matrix $|\gamma_{i\alpha}|$,(i=1,....,m;α =1,....,r) is called the dimensional matrix.

The product π_ρ,

$$\pi_\rho = [(z_1{}^{k_{\rho 1}} \cdots (z_m{}^{k_{\rho m}}] \tag{2.91}$$

$$\rho = 1, \ldots., n)$$

has a wide use in a number of engineering applications of dimensional analysis.

The functions $(\pi_\rho : \rho = 1, \ldots., n)$ defined in Eq.(2.91) are absolute invariants under a r-parameter dimensional group of transformations, if and only if the k's satisfy the system

$$\sum_{i=I}^{m} k_{\rho i} \gamma_{i\alpha} = 0 \quad (\alpha = 1, \ldots, r) \tag{2.92}$$

It is interesting to note that the absolute invariant functions or products are non-dimensional.

For clarification of the above concepts, consider the problem of deflection, δ, of a structure of a given shape under dynamic loading. δ would depend on the size of the structure which can be represented by a characteristic length L, modulus of elasticity E, mass density ρ_0, frequency Ω of the application of loading, amplitudes F and $\tilde{M}$ of the applied forces and moments, respectively. The dimensional group of transformations Γ_0 can be written as

$$\begin{aligned}
\Gamma_0 : \quad \bar{\delta} &= a_M^0 a_L^1 a_T^0 \delta \\
\bar{L} &= a_M^0 a_L^1 a_T^0 L \\
\bar{E} &= a_M^1 a_L^{-1} a_T^{-2} E \\
\bar{\rho} &= a_M^1 a_L^{-3} a_T^0 \rho_0 \\
\bar{\Omega} &= a_M^0 a_L^0 a_T^{-1} \Omega \\
\bar{F} &= a_M^1 a_L^1 a_T^{-2} F \\
\bar{\tilde{M}} &= a_M^1 a_L^2 a_T^{-2} \tilde{M}
\end{aligned} \tag{2.93}$$

The dimensional matrix becomes

	δ	L	E	ρ_0	Ω	F	$\tilde{M}$
a_M	0	0	1	1	0	1	1
a_L	1	1	-1	-3	0	1	2
a_T	0	0	-2	0	-1	-2	-2

Based on traditional dimensional analysis [18], the following relationship can be deduced:

$$\frac{\delta}{L} = \phi\left(\frac{F}{EL^2}, \frac{\tilde{M}}{EL^3}, \frac{\Omega L\rho_0^{1/2}}{E^{1/2}}\right) \tag{2.94}$$

where the quantities

$$\frac{\delta}{L}\ ;\ F/(EL^2)\ ,\ \tilde{M}/EL^3\ and\ \Omega L\rho_0^{1/2}/E^{1/2}$$

are the non-dimensional π terms. It can be easily verified that the π terms are absolutely invariant under the dimensional group of transformations,Γ_0. For example,

$$\frac{\bar{F}}{\bar{E}\bar{L}^2} = \frac{a_M a_L a_T^{-2} F}{a_M a_L^{-1} a_T^{-2} E a_M^0 a_L^2 a_T^0 L^2}$$

$$= \frac{F}{EL^2} \tag{2.95}$$

The affine group of trnsformations can be expressed as

$$\bar{Z}_i = A_i Z_i \qquad (i = 1, \ldots.m) \tag{2.96}$$

where A_i are the parameters of the group. The affine groups are utilized in the Hellums-Churchill procedure discussed in chapter 3 and 4.

The one-parameter dimensional group can be thought of as subsumed under the affine group,i.e.,

$$\bar{Z}_i = a^{\alpha_i} Z_i \tag{2.97}$$

where $A_i = a^{\alpha_i}$.

For this reason, the affine group defined in Eq.(2.96) can be considered somewhat more general than the one-dimensional group, Eq.(2.97).

2.12 Summary

In this chapter, the concepts of the continuous transformation groups were discussed in detail. The ideas of infinitesimal group of transformations and extended groups were introduced. The important concept of invariance of functions and differential equations under infinitesimal groups was elaborated. It was also shown that invariance of differential equations under an infinitesimal group of transformations led to a reduction in the number of independent variables. The notion of characteristic function as an invariant surface condition was introduced in this chapter. The invariant representation expressed in terms of the characteristic functions, offers simplification and elegance in derivation based on the deductive methods of invariance analysis as will be seen in the later chapters.

REFERENCES

[1] Lie, S., Math. Annalen,Vol.8,p.220 (1875).

[2] Lie, S. and Engel, F.,Theorie der Transformations- gruppen,Vols. 1-3,Teubner, Leipzig (1890).

[3] Lie, S. and Scheffers, G.,Vorlesungen über Differential-gleichungen mit bekannten infinitesimalen Transformationen, Teubner, Leipzig (1891).

[4] Cohen, A., Intro. to Lie Theory of One-parameter groups, Heath, New York (1931).

[5] Campbell, J. E., Introductory Treatise on Lie's Theory of Finite Continuous Transformation Groups, Oxford University Press (1963).

[6] Eisenhart, L.P.,Continuous Group of Transformations, Dover (1933).

[7] Ovsiannikov, L.V., Group Analysis in Differential Equations, Academic Press (English translation edited by W.F.Ames 1982).

[8] Bluman, G.W. and Cole, J.D.,Similarity Meth. for Dif. Eqs., Springer-Verlag, New York (1974).

[9] Birkhoff, G.,Hydrodynamics, Princeton University Press (1950).

[10] Morgan, A.J.A., "Reduction by One of the Number of Independent Variables in Some Systems of Partial Differential Equations", Quar. Appl. Math.,Vol.3,pp.250-259 (1952).

[11] Hansen, A.G., Similarity Analyses of Boundary Value Problems in Engineering, Prentice-Hall Inc. (1964).

[12] Na, T.Y., Abbott, D.E. and Hansen, A.G., "Similarity Analysis of Partial Differential Equations", Rept. NAS 8-20065, The University of Michigan, Dearborn, Michigan (1967).

[13] Na, T.Y. and Hansen, A.G., "Similarity Analysis of Differential Equations by Lie Group", J. of the Franklin Institute, Vol.6,p.292 (1971).

[14] Ames, W.F., Nonlinear Partial Differential Equations in Engineering, Vol.1, Academic Press Inc., New York (1965).

[15] Ames, W. F.,Nonlinear Partial Differential Equations in Engineering, Vol.2, Academic Press , New York (1972).

[16] Sneddon, I.N., Elements of Partial Differential Equations, McGraw-Hill, New York (1957).

[17] Ince, E.L.,Ordinary Differential Equations, Dover (1956).

[18] Baker, W.E., Westine, P.S. and Dodge, F.T., Similarity Methods in Engineering Dynamics, Spartan Books (1973).

[19] Mueller, E.A. and Matschat, K., Uber das Auffinden von Ahnlichkeitslosungen Partieller Differential Gleichungssysteme unter Benutzung von Transformations Gruppen, mit Anwendengunen auf Probleme der Stromungsphysik, Akademie-Verlag, Berlin (1962).

Chapter 3

A SURVEY OF METHODS FOR DETERMINING SIMILARITY TRANSFORMATIONS

3.0 Introduction

In this chapter, different methods for determining similarity transformations of partial differential equations will be discussed. A similarity transformation reduces the number of independent variables in the partial differential equations. The transformed system of equations and auxiliary conditions is known as a "similarity representation".

Direct methods of similarity analysis do not invoke invariance under a group of transformations. They are fairly straightforward and simple to apply. The separation of variables method of Abbott and Kline [1], and the method of dimensional analysis fall into the category of direct methods.

The group-theoretic methods on the other hand are more recent, and are mathematically elegant. The underlying basis in any group-theoretic method is that of invariance. Birkhoff[2] and Morgan's[3] method involves the use of an "assumed group of transformations" at the outset of the analysis. Hellums and Churchill's[4] procedure is essentially similar to the Birkhoff-Morgan method, however, the use of mass, length and time as fundamental dimensions is implied in the procedure. Consequently a non-dimensional similarity representation is obtained.

The deductive methods of similarity analysis start out with a general group of transformations. The equations and boundary conditions are rendered invariant under the general group, and similarity solutions are subsequently derived on a systematic basis. Deductive group methods are further classified into (a) finite group method and (b) infinitesimal group method. Moran and Gaggioli[5] have developed a deductive procedure based on finite group of transformations. The infinitesimal group method of Bluman and Cole[6] starts out with a general infinitesimal group of transformations and systematically deduces the similarity transformations. The characteristic-function method of Na and Hansen[7] is, again, based upon an infinitesimal group of transformations. However, the introduction of the characteristic function renders the subsequent mathematical description in terms of a single dependent variable. Since the finite group is generated by the infinitesimal group, it is adequate to use the infinitesimal group methods for the systematic derivation of similarity solutions.

3.1 Direct Methods

(a) Separation of Variables Method of Abbott and Kline[1]

Consider the linear heat equation

$$\frac{\partial^2 u}{\partial y^2} = \frac{\partial u}{\partial t} \tag{3.1}$$

The question we ask is whether a transformation of variables exists which reduces the number of independent variables in Eq.(3.1) from two to one? The dependent variable u would then transform to a function of a new independent variable ς alone, such that u(x,t) $\rightarrow$ $u(\varsigma)$ where ς=$\varsigma(y,t)$. The resulting equation would be second-order in terms of ς.

In the separation of variables method, the transformation is assumed as

$$u = b \cdot g(t) \cdot f(\varsigma) \tag{3.2}$$

where $\varsigma = ay/t^m$; a , b and m are constants that will be subsequently determined.

By the chain rule of calculus, we can write the partial derivatives as

$$\frac{\partial u}{\partial y} = \frac{\partial u}{\partial \varsigma} \cdot \frac{\partial \varsigma}{\partial y} = b \cdot g(t) \cdot \frac{d\,f}{d\varsigma} \cdot \frac{a}{t^m} \tag{3.3}$$

$$\frac{\partial^2 u}{\partial y^2} = \frac{\partial}{\partial y}[\frac{\partial u}{\partial y}] = \frac{\partial}{\partial \varsigma}[\frac{\partial u}{\partial y}] \cdot \frac{\partial \varsigma}{\partial y} = b \cdot g(t) \cdot \frac{d^2 f}{d\varsigma^2} \cdot \frac{a^2}{t^{2m}} \tag{3.4}$$

Similarly,

$$\frac{\partial u}{\partial t} = b \cdot \frac{dg}{dt} \cdot f(\varsigma) + b \cdot g(t) \cdot \frac{d\,f}{d\varsigma} \cdot \frac{d\varsigma}{dt}$$

$$= b \cdot [g'f - \frac{m\varsigma g f'}{t}] \tag{3.5}$$

Substituting Eqs.(3.3) to (3.5) into Eq. (3.1), we obtain the following transformed equation

$$bgf'' \frac{a^2}{t^{2m}} = b[g'f - \frac{m\varsigma g f'}{t}] \tag{3.6}$$

Dividing throughout by $b \cdot g \cdot f$ and rearranging,

$$(\frac{g'}{g})t^{2m} = (\frac{f''}{f})a^2 + \frac{m\varsigma}{t^{1-2m}}(\frac{f'}{f}) \tag{3.7}$$

An examination of Eq.(3.7) shows that if t^{1-2m}=constant, that is 1-2m=0, then the right hand side is a function of ς alone.

If the left hand side is equated to a constant λ such that $g't/g$=λ, then

$$g(t) = g_0 t^\lambda \tag{3.8}$$

where g_0 is another constant. Thus the required similarity transformation is

$$u = bg_0 t^\lambda f(\frac{ay}{\sqrt{t}}) \tag{3.9}$$

which is a well-known result.

The transformed ordinary differential equation can now be written as

$$a^2 f'' + \frac{1}{2}\varsigma f' - \lambda f = 0 \tag{3.10}$$

(B) The Method of Dimensional Analysis

The application of the method of dimensional analysis for finding similarity transformations has been nicely demonstrated by Sedov[8]. Generalizations to the method have been proposed by Moran[9] and Morrison[10]. In this section, we will use the dimensional analysis procedure as suggested by Moran and Morrisson, which we shall refer to as the "Modified Dimensional Analysis".

The success of dimensional method depends on the proper identification of the physical parameters and variables that go into the description of a physical problem. As an illustration, we will consider the following boundary value problem commonly known as the Rayleigh Flow Problem. An infinite plate is immersed in an incompressible fluid at rest. The plate is suddenly accelerated, so that it moves parallel to itself at a constant velocity, U_0. Let u be the fluid velocity in the x-direction, v and w the velocities in the y and z directions, respectively. From physical symmetry $v=w=0$, and the viscous-diffusion equation describing the flow can be written as

$$\nu \frac{\partial^2 u}{\partial y^2} = \frac{\partial u}{\partial t} \tag{3.11}$$

where ν is the kinematic viscosity, and t is the time.

The boundary and initial conditions are:

$$\begin{aligned} u(0,t) &= U_0 \quad t > 0 \\ u(y,0) &= 0 \quad y > 0 \\ u(\infty,t) &= 0 \end{aligned} \tag{3.12}$$

In the method of dimensional analysis used here, we will distinguish between lengths in different directions by assigning for each direction a separate dimension. By doing so, we will not lose the physical information that would be needed to discover the similarity transformation. The velocity in the x direction, u, can be expressed as

$$u = f(y, \nu, t, U_0) \tag{3.13}$$

The "dimensional matrix" can now be written as

	u	y	t	ν	u_0
M	0	0	0	0	0
L_x	1	0	0	0	1
L_y	0	1	0	2	0
T	-1	0	1	-1	-1
	a_1	a_2	a_3	a_4	a_5

The *rank* of the matrix is 3 and the *number of variables* are 5. Therefore, the number of *Pi terms* are 5-3=2.

The "dimensions" a_i in the dimensional matrix are defined by the realtionship

$$u^{a_1} y^{a_2} t^{a_3} \nu^{a_4} U_0^{a_5} = M^0 L_x{}^0 L_y{}^0 T^0 \tag{3.14}$$

where the equal sign means "dimensionally equal to".

The dimensional matrix is equivalent to the following system of equations:

$$\begin{aligned} a_1 + a_5 &= 0 \\ a_2 + 2a_4 &= 0 \\ -a_1 + a_3 - a_4 - a_5 &= 0 \end{aligned} \tag{3.15}$$

Solving Eqs.(3.15), and rewriting Eq.(3.14),

$$\left(\frac{u}{U_0}\right)^{a_1} \left[\frac{y}{(\nu t)^{1/2}}\right]^{a_2} = M^0 L_x{}^0 L_y{}^0 T^0 \tag{3.16}$$

where, again, the equal sign means 'dimensionally equal to'.

The Pi-terms are therefore

$$\pi_1 = \frac{u}{U_0} \quad ; \quad \pi_2 = \frac{y}{(\nu t)^{1/2}} \tag{3.17}$$

The similarity transformation can now be written as

$$\pi_1 = f(\pi_2) \tag{3.18}$$

or

$$\frac{u}{U_0} = f\left[\frac{y}{(\nu t)^{1/2}}\right] \tag{3.19}$$

As an exercise, it would be worthwhile for the reader to repeat the above example by using $L = L_x = L_y$, as is the case in "traditional" dimensional analysis. It will be discovered that a similarity transformation will not result. The reason lies in the fact that, physically speaking, u and U_0 are measured in the x direction while ν is defined in terms of the y direction. However, if the distinction between L_x and L_y is not made, some physical information is lost resulting in the failure of the method. We therefore recommend that such problems be analyzed by assigning separate dimensions to different directions, as is done in the Modified Dimensional Analysis proposed by Morrisson[10] and by Moran[11].

3.2 Group-Theoretic Methods

We have seen in chapter 2, that the invocation of invariance of a partial differential system under a group of transformations would lead to a reduction in the number of independent variables. The concept of invariance plays a key role in the mathematical formulation of the group-theoretic procedures. The main advantage of using group-theoretic procedures is that they are systematic. Invariance can be invoked either by using an "assumed" group, or by starting out with general groups and then applying deductive procedures.

(a) Birkhoff and Morgan's Method

This method was the first application of Lie's theories of continous transformation groups to similarity analysis. In this method, a group of transformations is defined a priori, and invariance of the partial differential equation is then invoked. Two groups, namely the linear and the spiral, are used here to illustrate the Birkhoff-Morgan method. Although the choice of assumed groups of transformations limits the generality of the results, the process for obtaining similarity solutions is clearly demonstrated. Furthermore, the majority of the problems of engineering interest are covered by the linear and spiral groups of transformations.

Consider again, the heat equation

$$\frac{\partial u}{\partial t} - \nu\frac{\partial^2 u}{\partial y^2} = 0 \tag{3.11}$$

subject to the auxiliary conditions

$$\begin{aligned} &u(y,0) = 0 \quad y > 0 \\ &u(0,t) = U(t) \quad t > 0 \\ &u(\infty,t) = 0 \end{aligned} \tag{3.20}$$

Consider the linear group of transformations:

$$G: \quad t = A^{\alpha_1}\bar{t} \ ; \ y = A^{\alpha_2}\bar{y} \ ; \ u = A^{\alpha_3}\bar{u} \tag{3.21}$$

where A is the parameter of transformations and α_1, α_2 and α_3 are constants which are as yet to be determined. Invariance of Eq.(3.11) under the above group gives:

$$\begin{aligned} &\frac{\partial u}{\partial t} - \nu\frac{\partial^2 u}{\partial y^2} \\ &= \frac{A^{\alpha_3}}{A^{\alpha_1}}\left(\frac{\partial \bar{u}}{\partial \bar{t}}\right) - \frac{A^{\alpha_3}}{A^{2\alpha_2}}\left(\nu\frac{\partial^2 \bar{u}}{\partial \bar{y}^2}\right) = 0 \end{aligned} \tag{3.22}$$

For Eq.(3.22) to be constant conformally invariant, i.e.,

$$\frac{\partial u}{\partial t} - \nu\frac{\partial^2 u}{\partial y^2} = F(A)[\frac{\partial \bar{u}}{\partial \bar{t}} - \nu\frac{\partial^2 \bar{u}}{\partial \bar{y}^2}]$$

the powers of A should be the same. Therefore,

$$\alpha_3 - \alpha_1 = \alpha_3 - 2\alpha_2 \tag{3.23}$$

or,

$$\alpha_2 = \frac{\alpha_1}{2} \tag{3.24}$$

Thus, the conditions set forth in section 2.6 are satisfied according to Morgan's theorems, and the partial differential equation can be expressed in terms of (m+n-1) absolute invariants. In the present example, there are two independent and one dependent variables, so that the heat equation can be expressed in terms of two absolute invariants. To find these invariants, it is necessary to eliminate the parameter of transformation, A, from Eq.(3.21). In other words, we are required to find the r and s such that

$$yt^r = \bar{y}\bar{t}^r \tag{3.25}$$

and

$$ut^s = \bar{u}\bar{t}^s$$

It can be easily seen that

$$r = -\frac{\alpha_2}{\alpha_1} = -\frac{1}{2} \quad ; \quad s = -\frac{\alpha_3}{\alpha_1} = -\delta$$

The similarity transformation can be written as

$$\frac{u}{t^\delta} = F(\frac{y}{t^{1/2}}) \tag{3.26}$$

It should be ascertained whether or not the auxiliary conditions are invariant under the group defined by Eq.(3.21). The auxiliary conditions u(y,0)=0 and $u(\infty$,t)=0 combine together as one boundary condition in the similarity coordinate, ς, such that F(∞)=0. Using Eq.(3.26) and transforming the auxiliary condition u(0,t)= U(t) to the similarity coordinate, we obtain the following:

$$u(0,t) = t\ F(0) = U(t) \tag{3.27}$$

Therefore, when $U(t) = U_0 t^\delta$, where $U_0 = F(0)$, all of the auxiliary conditions would be invariant under Eq.(3.21).

If we use the following boundary condition:

$$F(t) = U_0\, sin^{-1}(t) \tag{3.28}$$

then the auxiliary conditions will not be completely invariant, and a similarity transformation will not exist. However,by using superposition of similarity solutions, it may be possible to obtain the required solution. Such methods are discussed in chapter 6 of this book.

The similarity representation for the heat equation for a variation of $U(t)$ defined by $U(t)=U_0 t^\delta$, is

$$F'' + \frac{1}{2}\varsigma F' - \delta F = 0 \tag{3.29}$$

with boundary conditions

$$F(0) = U_0 \quad ; \quad F(\infty) = 0$$

Consider now the spiral group of transformation defined by

$$t = \bar{t} + \beta_1 a \ ; \ y = e^{\beta_2 a}\bar{y} \ ; \ u = e^{\beta_3 a}\bar{u} \tag{3.30}$$

Under this group, the invariant heat equation which takes the form

$$\frac{\partial u}{\partial t} - \nu\frac{\partial^2 u}{\partial y^2} = e^{\beta_3 a}\left(\frac{\partial \bar{u}}{\partial \bar{t}}\right) - \nu\frac{e^{\beta_3 a}}{e^{2\beta_2 a}}\left(\frac{\partial^2 \bar{u}}{\partial \bar{y}^2}\right) = 0$$

is constant conformally invariant under Eq.(3.30) if

$$\beta_3 = \beta_3 - 2\beta_2 \tag{3.31}$$

or

$$\beta_2 = 0$$

The absolute invariants are

$$\eta = y \quad ; \quad F(\eta) = \frac{u}{e^{\beta t}} \tag{3.32}$$

where

$$\beta = \frac{\beta_3}{\beta_1}$$

The boundary conditions become

$$F(0) = C \quad ; \quad F(\infty) = 0$$

The transformed heat equation is

$$\nu F'' - \beta F' = 0 \tag{3.33}$$

The solution to Eq.(3.33) with boundary conditions $F(0)=C$ and $F(\infty)=0$ is

$$u(y,t) = C\ exp(\beta t - \sqrt{\frac{\beta}{\nu}}y) \tag{3.34}$$

From this solution it is clear that the similarity solution will exist only if the initial condition is

$$u(y,0) = C\, e^{-\sqrt{\beta/\nu}y}$$

(b) The Hellums-Churchill Procedure

Many useful similarity solutions cannot be discovered by the use of traditional dimensional analysis. The use of simple affine transformations would give rise to new results even in the relatively well-trodden fields of fluid mechanics and heat transfer. Hellums-Churchill[4] proposed two extensions to Birkhoff's method of search for symmetric solutions by using the more valuable affine group of transformations. The first extension is that the problem of finding the minimum parametric description can directly be related to the problem of finding the minimum description in terms of the independent variables. In other instances, a non-dimensional representation can be obtained without achieving a reduction in the number of independent variables. The second extension proposed is that the method can be applied to yield the classes of functions which admit the possibility of a similarity transformation.

The Hellums-Churchill method implies the routine selection of mass, length and time as fundamental dimensions, and therefore, the method is suitable for physical problems. Since the transformation used is an assumed affine transformation group, the results while more general than those obtained by traditional dimensional analysis, are still restrictive.

We will now illustrate the use of the Hellums and Churchill procedure by applying it to the problem of one-dimensional heat conduction of a semi-infinite slab

$$\alpha \frac{\partial^2 T}{\partial y^2} = \frac{\partial T}{\partial t} \tag{3.35}$$

with the auxiliary conditions

$$T(0,t) = T_s \;\; ; \;\; T(\infty,t) = 0 \;\; ; \;\; T(y,0) = 0 \tag{3.36}$$

The method of analysis consists of the following steps:

(i) The variables in the problem description are rendered dimensionless, by introducing arbitrary reference variables. Therefore,

$$\bar{T} = \frac{T - T_A}{T_0} \;\; ; \;\; \bar{y} = \frac{y}{y_0} \;\; ; \;\; \frac{t}{t_0}$$

where T_0, T_A, y_0 and t_0 are the arbitrary reference variables, to be determined so that a minimum parametric description would result.

(ii) The equations and auxiliary conditions are rendered dimensionless using transformations,Eq.(3.35), as follows:

$$\frac{\partial \bar{T}}{\partial \bar{t}} = \left(\frac{\alpha t_0}{{y_0}^2}\right)\frac{\partial^2 \bar{T}}{\partial \bar{y}^2}$$

$$\bar{T}(0,\bar{t}) = \frac{T_S - T_A}{T_0} \tag{3.37}$$

$$\bar{T}(\bar{y},0) = -\frac{T_A}{T_0}$$

(iii) The reference varaibles are determined by obtaining a minimum parametric description. The problem description can be expressed as

$$\frac{T - T_A}{T_0} = f\left(\frac{y}{y_0}, \frac{t}{t_0}, \frac{\alpha t_0}{{y_0}^2}, \frac{T_S - T_A}{T_0}, \frac{T_A}{T_0}\right) \tag{3.38}$$

For minimum parametric description, we set

$$\frac{\alpha t_0}{{y_0}^2} = 1 \ ; \quad \frac{T_s - T_A}{T_0} = 0 \ ; \quad \frac{T_A}{T_0} = 1$$

Therefore,

$$\alpha t_0 = y_0^2 \ ; \quad T_A = T_S \ ; \quad T_0 = T_A = T_S \tag{3.39}$$

Therefore, the minimum description is

$$\frac{T - T_S}{T_S} = f\left(\frac{y}{y_0}, \frac{\alpha t}{{y_0}^2}\right) \tag{3.40}$$

(iv) Since y_0 does not appear in the original description, it can be suitably eliminated to give the following

$$\frac{T - T_S}{T_S} = f\left(\frac{y}{\sqrt{\alpha t}}\right) \tag{3.41}$$

Eq.(3.41) is the required similarity transformation.

If the boundary condition at y=0 is replaced by a constant heat flux condition, then

$$q(0,t) = q_0 = K\frac{\partial T(0,t)}{\partial y}$$

The description of the problem becomes

$$\frac{T - T_A}{T_0} = \phi\left(\frac{y}{y_0}, \frac{t}{t_0}, \frac{\alpha t_0}{y_0^2}, \frac{K T_0}{q_0 y_0}, \frac{T_A}{T_0}\right)$$

The minimum parametric description gives

$$\alpha t_0 = {y_0}^2 \quad ; \quad T_0 = \frac{q_0 y_0}{K} \quad ; \quad T_A = 0 \tag{3.42}$$

Therefore, the similarity transformation is given by

$$\frac{K\ T(y,t)}{q_0\sqrt{\alpha t}} = g\left(\frac{y}{\sqrt{\alpha t}}\right) \tag{3.43}$$

A group-theoretic viewpoint of the Hellums-Churchill procedure is discussed by Moran[11] and Seshadri[12]. The use of affine groups, as is implied in the Hellums-Churchill method, could lead to results that are not obtainable by using traditional dimensional analysis.

(c) Deductive Similarity Analysis : Moran and Gaggioli's Method

The main drawback of methods which use an assumed group of transformation at the outset of the analysis is that the resulting solutions are restrictive. Therefore, if an invariant solution cannot be discovered, one should not conclude that similarity solutions do not exist. Recourse to deductive methods of analysis using general groups of transformations could systematically lead to a number of similarity solutions.

Consider again the heat equation given in Eq.(3.11). Instead of proceeding with an assumed group of transformations as was done in the earlier sections, a general one-parameter finite group of transformations is introduced as follows:

$$\begin{aligned} G: \quad \bar{y} &= f^y(y,t;a) \\ \bar{t} &= f^t(y,t;a) \\ \bar{u} &= f^u(u;a) \end{aligned} \tag{3.44}$$

The group G can be enlarged using the chain rule of differentiation as follows:

$$\begin{aligned} \frac{\partial \bar{u}}{\partial \bar{t}} &= \left(\frac{\partial \bar{u}}{\partial u}\frac{\partial t}{\partial \bar{t}}\right)\frac{\partial u}{\partial t} \\ &+\left(\frac{\partial \bar{u}}{\partial u}\frac{\partial y}{\partial \bar{t}}\right)\frac{\partial u}{\partial y} \end{aligned} \tag{3.45}$$

$$\begin{aligned} \frac{\partial^2 \bar{u}}{\partial \bar{y}^2} &= \left[\frac{\partial \bar{u}}{\partial u}\left(\frac{\partial y}{\partial \bar{y}}\right)^2\right]\frac{\partial^2 u}{\partial y^2} + \left(\frac{\partial \bar{u}}{\partial u}\frac{\partial^2 t}{\partial \bar{y}^2}\right)\frac{\partial u}{\partial t} \\ &+\left(\frac{\partial \bar{u}}{\partial u}\frac{\partial^2 y}{\partial \bar{y}^2}\right)\frac{\partial u}{\partial y} + \left[\frac{\partial \bar{u}}{\partial u}\left(\frac{\partial t}{\partial \bar{y}}\right)^2\right]\frac{\partial^2 u}{\partial t^2} \\ &+2\left(\frac{\partial \bar{u}}{\partial u}\frac{\partial y}{\partial \bar{y}}\frac{\partial t}{\partial \bar{y}}\right)\frac{\partial^2 u}{\partial y \partial t} + \left[\frac{\partial^2 \bar{u}}{\partial u^2}\left(\frac{\partial y}{\partial \bar{y}}\right)^2\right]\left(\frac{\partial u}{\partial y}\right)^2 \end{aligned}$$

$$+[\frac{\partial^2 \bar{u}}{\partial u^2}(\frac{\partial t}{\partial \bar{y}})^2](\frac{\partial u}{\partial t})^2$$

$$+2(\frac{\partial^2 \bar{u}}{\partial u^2}\frac{\partial t}{\partial \bar{y}}\frac{\partial y}{\partial \bar{y}})\frac{\partial u}{\partial t}\frac{\partial u}{\partial y} \tag{3.46}$$

The first condition to be satisfied by the general one-parameter transformation group is the requirement that the given partial differential equation, Eq.(3.1), be conformally invariant. According to Morgan's theorem, the conformal invariance of Eq.(3.1) requires that the following equation be satisfied:

$$\frac{\partial \bar{u}}{\partial \bar{t}} - \nu\frac{\partial^2 \bar{u}}{\partial \bar{y}^2}$$

$$= F(t, y, u, \frac{\partial u}{\partial t}, ...; a)[\frac{\partial u}{\partial t} - \nu\frac{\partial^2 u}{\partial y^2}] \tag{3.47}$$

Substitution of Eqs.(3.45) and (3.46) into Eq.(3.47) gives:

$$[\frac{\partial \bar{u}}{\partial u}(\frac{\partial t}{\partial \bar{t}} - \nu\frac{\partial^2 t}{\partial \bar{y}^2})]\frac{\partial u}{\partial t} - \nu[\frac{\partial \bar{u}}{\partial u}(\frac{\partial y}{\partial \bar{y}})^2]\frac{\partial^2 u}{\partial y^2} + R$$

$$= F[\frac{\partial u}{\partial t} - \nu\frac{\partial^2 u}{\partial y^2}] \tag{3.48}$$

where

$$R = [\frac{\partial \bar{u}}{\partial u}(\frac{\partial y}{\partial \bar{t}} - \nu\frac{\partial^2 y}{\partial \bar{y}^2})]\frac{\partial u}{\partial y} - \nu[\frac{\partial \bar{u}}{\partial u}(\frac{\partial t}{\partial \bar{y}})^2]\frac{\partial^2 u}{\partial t^2}$$

$$-2\nu(\frac{\partial \bar{u}}{\partial u}\frac{\partial y}{\partial \bar{y}}\frac{\partial t}{\partial \bar{y}})\frac{\partial^2 u}{\partial y \partial t} - \nu[\frac{\partial^2 \bar{u}}{\partial u^2}(\frac{\partial t}{\partial \bar{y}})^2](\frac{\partial u}{\partial t})^2$$

$$-\nu[\frac{\partial^2 \bar{u}}{\partial u^2}(\frac{\partial y}{\partial \bar{y}})^2](\frac{\partial u}{\partial y})^2$$

$$-2\nu[\frac{\partial^2 \bar{u}}{\partial u^2}\frac{\partial t}{\partial \bar{y}}\frac{\partial y}{\partial \bar{y}}]\frac{\partial u}{\partial t}\frac{\partial u}{\partial y}$$

It is seen that conformal invariance will result if, simultaneously, R=0 and

$$\frac{\partial t}{\partial \bar{t}} - \nu\frac{\partial^2 t}{\partial \bar{y}^2} = (\frac{\partial y}{\partial \bar{y}})^2 \tag{3.49}$$

For R to vanish, it is sufficient that the coefficients of the derivatives

$$\frac{\partial u}{\partial y}, \quad \frac{\partial^2 u}{\partial y^2}, \quad \frac{\partial^2 u}{\partial y \partial t} \quad and \quad \frac{\partial u}{\partial t}$$

identically vanish, as follows:

$$\frac{\partial \bar{u}}{\partial u}(\frac{\partial y}{\partial \bar{t}} - \nu\frac{\partial^2 y}{\partial \bar{y}^2}) \equiv 0$$

$$\frac{\partial \bar{u}}{\partial u}(\frac{\partial t}{\partial \bar{y}})^2 \equiv 0$$

$$\frac{\partial \bar{u}}{\partial u}(\frac{\partial y}{\partial \bar{y}}\frac{\partial t}{\partial \bar{y}}) \equiv 0$$

$$\frac{\partial^2 \bar{u}}{\partial u^2}(\frac{\partial t}{\partial \bar{y}})^2 \equiv 0 \tag{3.50}$$

$$\frac{\partial^2 \bar{u}}{\partial u^2}(\frac{\partial y}{\partial \bar{y}})^2 \equiv 0$$

$$\frac{\partial^2 \bar{u}}{\partial u^2}(\frac{\partial t}{\partial \bar{y}}\frac{\partial y}{\partial \bar{y}}) \equiv 0$$

Now,since the derivatives

$$\frac{\partial \bar{u}}{\partial u}, \frac{\partial \bar{y}}{\partial y}, \frac{\partial \bar{t}}{\partial t}, \frac{\partial y}{\partial \bar{y}} \; and \; \frac{\partial t}{\partial \bar{t}}$$

do not vanish, Eq.(3.50) is satisfied if

$$\frac{\partial y}{\partial \bar{t}} - \nu \frac{\partial^2 y}{\partial \bar{y}^2} \equiv 0 \tag{3.51}$$

$$\frac{\partial t}{\partial \bar{y}} \equiv 0 \tag{3.52}$$

$$\frac{\partial^2 \bar{u}}{\partial u^2} \equiv 0 \tag{3.53}$$

Eq.(3.52) indicates that Eq.(3.49) can be simplified to

$$\frac{\partial t}{\partial \bar{t}} = (\frac{\partial y}{\partial \bar{y}})^2 \tag{3.54}$$

Multiplying Eq.(3.51) by $2(\partial y/\partial \bar{y})$ and making use of Eq.(3.54), we get:

$$2(\frac{\partial y}{\partial \bar{y}})(\frac{\partial y}{\partial \bar{t}}) - \nu \frac{\partial^2 t}{\partial \bar{y} \partial \bar{t}} \equiv 0 \tag{3.55}$$

The second term is seen to be zero from Eq.(3.52). Since $\partial y/\partial \bar{y}$ is not zero, Eq.(3.55) indicates that:

$$\frac{\partial y}{\partial \bar{t}} = 0 \tag{3.56}$$

Eq.(3.51) therefore becomes

$$\frac{\partial^2 y}{\partial \bar{y}^2} \equiv 0 \tag{3.57}$$

Eqs.(3.56) and (3.57) show that

$$y = K_1(a)\bar{y} + K_2(a)$$

or,

$$\bar{y} = c_1(a)\ y + c_2(a) \tag{3.58}$$

Substituting y from Eq.(3.58) into Eq.(3.54) and integrating, the following relation between $\bar{t}$ and t is obtained:

$$\bar{t} = [c_1(a)]^2 t + c_3(a) \tag{3.59}$$

Eq.(3.53) yields

$$\bar{u} = c_4(a)u + c_3(a) \tag{3.60}$$

Thus, the diffusion equation, Eq.(3.1), is conformally invariant under the group of transformations,

$$G: \quad \bar{y} = f^y(y;a) = c_1(a)y + c_2(a)$$

$$\bar{t} = f^t(t;a) = [c_1(a)]^2 t + c_3(a)$$

$$\bar{u} = f_u(u;a) = c_4(a)u + c_5(a) \tag{3.61}$$

Based on Morgan's theorem, if a given partial differential equation transforms conformally under a group of transformations, then it can be expressed in terms of the functionally independent invariants of this group. Before the invariants are found, however, additional restrictions will be placed on the function f, by the requirement that each of the auxiliary conditions (i.e., the boundary connitions and the initial conditions) be satisfied by solutions invariant under the transformation group G.

Let us denote an invariant solution by $I(z^1, z^2)$,i.e.,for all values of the group parameter a,

$$f_u(u;a) = I[f^y(y;a), f^t(t;a)] \tag{3.62}$$

Since this is an invariant solution, it can be written as

$$u = I(y,t) \tag{3.63}$$

For group G, defined in Eq.(3.61), Eq.(3.62) gives

$$c_4(a)u + c_5(a)$$

$$= I[c_1(a)y + c_2(a), (c_1(a))^2 t + c_3(a)] \tag{3.64}$$

Specializing the group to account for the auxiliary conditions as stated in Eq.(3.12), we get:

$$I[c_1(a)y + c_2(a), c_3(a)] = c_5(a)$$

$$I[c_2(a), c_1^2(a)t + c_3(a)] = c_4(a)U_0 + c_5(a)$$

$$I[\infty, c_1^2(a)t + c_3(a)] = c_5(a) \tag{3.65}$$

After considerable algebraic manipulations, the following can be found to hold:

$$c_2(a) = 0$$

$$c_3(a) = 0$$

$$c_4(a) = 1$$

$$c_5(a) = 0$$

Thus, to satisfy Morgan's sufficient condition for the existence of a similarity transformation of the partial differential equation, and to satisfy the necessary condition imposed by the auxiliary conditions, a transformation group must be of the form:

$$G: \quad \bar{y} = f^y(y; a) = c_1(a)y$$

$$\bar{t} = f^t(t; a) = [c_1(a)]^2 t$$

$$\bar{u} = f_u(u; a) = u \tag{3.66}$$

The two absolute invariants required can be established by Eq.(2.20), with

$$\frac{dy}{[f^y(y; a_0)]_a} = \frac{dt}{[f^t(t; a_0)]_a} = \frac{du}{[f_u(u; a_0)]_a} \tag{3.67}$$

where the subscript a in Eq.(3.67) represents partial differentiation with respect to a. Eq.(3.67) can also be written as

$$\frac{dy}{c_1'(a)y} = \frac{dt}{2c_1(a)c_1'(a)t} = \frac{du}{0}$$

The absolute invariants are therefore:

$$\eta = \frac{y}{2\sqrt{(\nu t)}} \quad and \quad F(\eta) = \frac{u}{U_0}$$

and the differential equation and its auxiliary conditions then become:

$$F_{\eta\eta} + 2\eta f_\eta = 0$$

with the boundary conditions:

$$F(\infty) = 0 \quad ; \quad F(0) = 1$$

The deductive group method of Gaggioli and Moran (1966) can therefore be seen to be very general. No specific form of the transformation group is assumed at the outset. Conformal invariance of the given partial differential equation under the group determines the transformation functions. The absolute invariants are then found, by a proper imposition of the auxiliary conditions. Thus, the trial and error nature of Birkhoff and Morgan's method is overcome.

Another approach was proposed by Moran and Gaggioli (1968a), in which the class of transformation groups introduced is of a somewhat more special, though still rather general form. Instead of introducing a very general class of groups, as in Eq.(3.44), a relatively special form, such as:

$$\bar{y} = c^{y}(a)y + K^{y}$$

$$\bar{t} = c^{t}(a)t + K^{y}$$

$$\bar{u} = c^{u}(a)u + K^{u}$$

is introduced in place of Eq.(3.44). However, the other steps remain unchanged. This method was extended for the multi-parameter groups and was successfully applied to the similarity solutions of three-dimensional boundary layer equations (Moran,Gaggioli and Scholten[13]) and the compressible boundary layer equations (Gaggioli and Moran[14]).

Other deductive group methods based on finite groups are discussed in detail by Ames[15].

(d) Infinitesimal Group Method (Bluman and Cole)

The deductive group method of Bluman and Cole[6] starts out with a general infinitesimal group of transformations. By invocation of invariance under the infinitesimal group the "determining equations" are derived. The determining equations are a set of linear differential equations, the solution of which gives the transformation function or the infinitesimals of the dependent and independent variables.

The heat equation $u_{yy} - u_t = 0$ is invariant under an infinitesimal group of transformations:

$$\begin{aligned} \bar{u} &= u + \epsilon U(y,t,u) + O(\epsilon^2) \\ \bar{y} &= y + \epsilon Y(y,t,u) + O(\epsilon^2) \\ \bar{t} &= t + \epsilon T(y,t,u) + O(\epsilon^2) \end{aligned} \tag{3.68}$$

if and only if

$$\bar{u}_{\bar{y}\bar{y}} - \bar{u}_{\bar{t}} = 0$$

For a boundary value problem, the boundary conditions would also be invariant. U, Y and T are the infinitesimals of the group of transformations.

As we have seen in section 2.9, the solution of the invariant surface equation

$$U(y,t,u) = Y(y,t,u)\frac{\partial u}{\partial x} + T(y,t,u)\frac{\partial u}{\partial t} \tag{3.69}$$

is the similarity solution. Bluman and Cole's technique differs from the method used by Na and Hansen in that the invariant surface is not identified with the characteristic function. Subsequently, the determining equations are expressed in terms of X, T and U, and their derivatives.

In order to evaluate the derivatives, we need the following transformations as has been shown in section 2.9:

$$\frac{\partial y}{\partial \bar{y}} = \frac{\partial}{\partial \bar{y}}[\bar{y} - \epsilon Y + O(\epsilon^2)]$$

$$= 1 - \epsilon[\frac{\partial Y}{\partial y}\frac{\partial y}{\partial \bar{y}} + \frac{\partial Y}{\partial u}\frac{\partial u}{\partial x}\frac{\partial x}{\partial \bar{x}}] + O(\epsilon^2)$$

$$= 1 - \epsilon[Y_y + Y_u u_x] + O(\epsilon^2) \tag{3.70a}$$

Similarly,

$$\frac{\partial y}{\partial \bar{t}} = -\epsilon[Y_t + Y_u u_t] + O(\epsilon^2) \tag{3.70b}$$

$$\frac{\partial t}{\partial \bar{t}} = 1 - \epsilon[T_t + T_u u_t] + O(\epsilon^2) \tag{3.70c}$$

$$\frac{\partial t}{\partial \bar{y}} = -\epsilon[T_y + T_u u_x] + O(\epsilon^2) \tag{3.70d}$$

If we recall that $\bar{u} = u + \epsilon U + O(\epsilon^2)$,

$$\frac{\partial \bar{u}}{\partial \bar{y}} = \frac{\partial \bar{u}}{\partial y}\frac{\partial y}{\partial \bar{y}} + \frac{\partial \bar{u}}{\partial t}\frac{\partial t}{\partial \bar{y}} \tag{3.71}$$

Substituting Eq.(3.70) into Eq.(3.71), we get

$$\frac{\partial \bar{u}}{\partial \bar{y}} = u_y + \epsilon[U_y + (U_u - Y_y)u_y - T_y u_t$$

$$- Y_u {u_y}^2 - T_u u_y u_t] + O(\epsilon^2) \tag{3.72}$$

Similarly,

$$\bar{u}_{\bar{y}\bar{y}} = u_{yy} + \epsilon[- T_{uu} u_y u_y u_t - Y_{uu} u_y u_y u_y$$

$$-2T_u u_y u_{yt} - (3Y_u + 2T_{yu})u_y u_t - T_u u_t u_t$$

$$+(U_{uu} - 2Y_{uy})u_y u_y - 2T_y u_{ty}$$

$$+(U_u - 2Y_y - T_{yy})u_t + (2U_{yu} - Y_{yy})u_y$$

$$+U_{yy}] + O(\epsilon^2) \tag{3.73}$$

$$\bar{u}_{\bar{t}} = u_t + \epsilon\big[- Y_u u_t u_y - T_u u_t u_t$$
$$+(U_u - T_t) u_t - Y_t u_y + U_t\big] + O(\epsilon^2) \tag{3.74}$$

substituting $u_{yy} = u_t$, and equating to zero the coefficients of

$$u_y u_{ty},\, u_y u_t,\, u_y u_y,\, u_{ty},\, u_t u_t,\, u_t,\, u_y$$

and u and the remaining terms in the expression of

$$\bar{u}_{\bar{y}\bar{y}} - \bar{u}_{\bar{t}} = 0$$

the determining equations can be obtained as follows:

$$u_y u_{ty}: \qquad T_u = 0 \tag{3.75a}$$

$$u_t u_y: \qquad Y_u = 0 \tag{3.75b}$$

$$u_y u_y: \qquad U_{uu} = 0 \tag{3.75c}$$

Therefore, solving Eq.(3.75a to d) we have

$$U(y,t,u) = f(y,t)u + g(y,t)$$
$$Y(y,t,u) = Y(y,t) \tag{3.76}$$
$$T(y,t,u) = T(y,t)$$

The remainder of the determining equations are

$$u_{ty} \;:\; T_y = 0 \;\; or \;\; T = T(t) \tag{3.77a}$$

$$u_t \;:\; 2Y_y - T'(t) = 0 \tag{3.77b}$$

$$u_y \;:\; Y_t - Y_{yy} + 2f_y = 0 \tag{3.77c}$$

$$u \;:\; f_{yy} - f_t = 0 \tag{3.77d}$$

$$u^0 \;:\; g_{yy} - g_t = 0 \tag{3.77e}$$

We will consider the subgroup for which $g(y,t) = 0$. Solving Eq.(3.77b) for y, we have

$$Y = \frac{yT'(t)}{2} + A(t) \tag{3.78}$$

where $A(t)$ is arbitrary. Substituting Eq.(3.78) into Eq.(3.77c) and then solving for f, we obtain

$$f = -\frac{y^2}{8}T''(t) - \frac{yA'(t)}{2} + B(t) \tag{3.79}$$

where $B(t)$ is arbitrary.

Substituting Eq.(3.79) into Eq.(3.77d), we have

$$-\frac{T''(t)}{4}+\frac{y^2T'''(t)}{8}+\frac{yA''(t)}{2}-B'(t)=0 \tag{3.80}$$

Solving Eq.(3.80), we can obtain the infinitesimals U, Y and T as:

$$Y = c_4 + 2c_1 t + c_5 y + 4c_2 yt$$

$$T = c_6 + 2c_5 t + 4c_2 t^2$$

$$U = -[2c_2 t + c_3 + c_1 y + c_2 y^2]u$$

c_1, c_2, c_3, c_4, c_5 and c_6 are six arbitrary parameters. All the parameters excepting c_2, individually represent transformations that can be obtained by inspectional procedures. c_4 represents translation invariance in y, c_6 translation in t, c_1 represents invariance under a Galilean transformation, and c_5 represents similitudinous invariance.

Similarity solution can now be derived by solving the invariant surface equation,Eq.(3.69), as

$$\frac{du}{U(y,t,u)}=\frac{dy}{Y(y,t,u)}=\frac{dt}{T(y,t,u)}$$

Specific similarity solutions are condidered in the next section.

(e) The Characteristic Function method (Na and Hansen)

The deductive group methods of Na and Hansen[7] and that of Bluman and Cole[6] are both based upon the invocation of invariance of the partial differential equations under an infinitesimal group of transformations. While Bluman and Cole determine the infinitesimals directly by solving the determining equations, Na and Hansen express the infinitesimals of the group in terms of a single function, W, called the "characteristic function". The procedure for finding the infinitesimals then reduces to the determination of the characteristic function. The latter procedure is convenient and more systematic, as will now be demonstrated by the analysis of the heat equation.

Using the notation

$$p_1=\frac{\partial u}{\partial t}, \; p_2=\frac{\partial u}{\partial y}, \; p_{11}=\frac{\partial^2 u}{\partial t^2}$$

$$p_{22}=\frac{\partial^2 u}{\partial y^2} \quad and \quad p_{12}=\frac{\partial^2 u}{\partial t\partial y}$$

the heat equation can be written as:

$$p_1 - p_{22} = 0 \tag{3.81}$$

We now define an infinitesimal group of transformations as follows:

$$
\begin{aligned}
\bar{t} &= t + \epsilon\xi(t, y, u) + O(\epsilon^2) \\
\bar{y} &= y + \epsilon\eta(t, y, u) + O(\epsilon^2) \\
\bar{u} &= u + \epsilon\varsigma(t, y, u) + O(\epsilon^2) \\
\bar{p_1} &= p_1 + \epsilon\pi_1(t, y, u, p_1, p_2) + O(\epsilon^2) \\
\bar{p_2} &= p_2 + \epsilon\pi_2(t, y, u, p_1, p_2) + O(\epsilon^2) \\
\bar{p_{22}} &= p_{22} + \epsilon\pi_{22}(t, y, u, p_1, p_2, p_{11}, p_{12}, p_{22}) + O(\epsilon^2)
\end{aligned}
\tag{3.82}
$$

where the transformation functions, $\xi, \eta, \varsigma, \pi_1, \pi_2$ and π_{22}, can be expressed in terms of the characteristic function, W, as:

$$
\begin{aligned}
\xi &= \frac{\partial W}{\partial p_1} \\
\eta &= \frac{\partial W}{\partial p_2} \\
\varsigma &= p_1 \frac{\partial W}{\partial p_1} + p_2 \frac{\partial W}{\partial p_2} - W \\
\pi_1 &= -\frac{\partial W}{\partial t} - p_1 \frac{\partial W}{\partial u} \\
\pi_2 &= -\frac{\partial W}{\partial y} - p_2 \frac{\partial W}{\partial u} \\
-\pi_{22} &= \frac{\partial^2 W}{\partial y^2} + 2p_2 \frac{\partial^2 W}{\partial y \partial u} + {p_2}^2 \frac{\partial^2 W}{\partial u^2} \\
&+2p_{12}\left(\frac{\partial^2 W}{\partial y \partial p_1} + p_2 \frac{\partial^2 W}{\partial u \partial p_1}\right) \\
&+2p_{22}\left(\frac{\partial^2 W}{\partial y \partial p_2} + p_2 \frac{\partial^2 W}{\partial u \partial p_2}\right) \\
&+{p_{12}}^2 \frac{\partial^2 W}{\partial {p_1}^2} + 2p_{12}p_{22} \frac{\partial^2 W}{\partial p_1 \partial p_2} \\
&+{p_{22}}^2 \frac{\partial^2 W}{\partial {p_2}^2} + p_{22} \frac{\partial W}{\partial u}
\end{aligned}
\tag{3.83}
$$

The characteristic function, W, is a function of t, y, u, p_1 and p_2.

Under the infinitesimal group of transformations, a partial differential equation $f=0$ will be invariant if $Uf=0$, i.e.,

$$
U(p_1 - p_{22}) = 0 \tag{3.84}
$$

or, writing the operation in its expanded form,

$$\xi\frac{\partial(\,)}{\partial t}+\eta\frac{\partial(\,)}{\partial y}+\varsigma\frac{\partial(\,)}{\partial u}+\pi_1\frac{\partial(\,)}{\partial p_1}$$

$$+\pi_2\frac{\partial(\,)}{\partial p_2}+\pi_{11}\frac{\partial(\,)}{\partial p_{11}}+\pi_{12}\frac{\partial(\,)}{\partial p_{12}}$$

$$+\pi_{22}\frac{\partial(\,)}{\partial p_{22}}=0 \tag{3.85}$$

where the parenthesis represents the quantity $p_1 - p_{22}$. Eq.(3.85) gives:

$$\pi_1-\pi_{22}=0 \tag{3.86}$$

which, upon substitution from Eq.(3.83), becomes:

$$-\frac{\partial W}{\partial t}-p_1\frac{\partial W}{\partial u}+\frac{\partial^2 W}{\partial y^2}+2p_2\frac{\partial^2 W}{\partial y\partial u}$$

$$+{p_2}^2\frac{\partial^2 W}{\partial u^2}+2p_{12}\frac{\partial^2 W}{\partial y\partial p_1}+2p_{12}p_2\frac{\partial^2 W}{\partial u\partial p_1}$$

$$+2p_1\frac{\partial^2 W}{\partial y\partial p_2}+p_1p_2\frac{\partial^2 W}{\partial u\partial p_2}+{p_{12}}^2\frac{\partial^2 W}{\partial {p_1}^2}$$

$$+2p_{12}p_1\frac{\partial^2 W}{\partial p_1\partial p_2}+{p_1}^2\frac{\partial^2 W}{\partial {p_1}^2}+p_1\frac{\partial W}{\partial u}=0 \tag{3.87}$$

Eq.(3.87) is used to solve for the characteristic function, $W(t,y,u,p_1,p_2)$.

Since W is not a function of p_{12}, the coefficients of the terms involving p_{12} and ${p_{12}}^2$ should be zero. By putting the coefficients of ${p_{12}}^2$ to zero, the characteristic function W is seen to be a linear function of p, i.e.,

$$W=W_1(t,y,u,p_2)+p_1\,W_2(t,y,u,p_2) \tag{3.88}$$

The coefficients of terms involving p_{12} then gives:

$$\frac{\partial W_2}{\partial y}+p_2\frac{\partial W_2}{\partial u}+p_1\frac{\partial W_2}{\partial p_2}=0 \tag{3.89}$$

Since W_2 is not a function of p_1, the coefficients of p_1 in Eq.(3.89) must be zero, which shows that W_2 is independent of p_2, i.e.,

$$W_2=W_2(t,u,y)$$

The remaining two terms in Eq.(3.89) then lead to the conclusion that W_2 is independent of both y and u and , as a result, the characteristic function now takes the form

$$W=W_1(t,y,u,p_2)+p_1\,W_2(t)$$

Putting this form of W into Eq.(3.87) and noting that both W_1 and W_2 are independent of p_1, Eq.(3.87) can be separated into three equations, corresponding to the coefficients of p_1^0,p_1^1 and p_1^2. We then get

$$p_1^0 \; : \quad -\frac{\partial W_1}{\partial t} + \frac{\partial^2 W_1}{\partial y^2} + 2p_2 \frac{\partial^2 W_1}{\partial y \partial u}$$

$$+ p_2^2 \frac{\partial^2 W_1}{\partial u^2} = 0$$

$$p_1^1 \; : \quad -\frac{\partial W_2}{\partial t} + 2\frac{\partial^2 W_1}{\partial y \partial p_2} + p_2 \frac{\partial^2 W_1}{\partial u \partial p_2} = 0$$

$$p_1^2 \; : \quad \frac{\partial^2 W_1}{\partial {p_2}^2} = 0 \qquad (3.90a, b, c)$$

From Eq.(3.90c),

$$W_1 = W_{11}(t, y, u) + p_2 W_{12}(t, y, u)$$

Putting this form of W_1 into Eq.(3.90b), we get:

$$-\frac{dW_2}{dt} + 2\frac{\partial W_{12}}{\partial y} + p_2 \frac{\partial W_2}{\partial u} = 0 \qquad (3.91)$$

Both W_2 and W_{12} are independent of p_2, the coefficient of p_2 in the third term should be zero, which means W_{12} is independent of u. Eq.(3.91) then becomes:

$$-\frac{dW_2}{dt} + 2\frac{\partial W_{12}}{\partial y} = 0 \qquad (3.92)$$

Since W_2 is a function of t only, Eq.(3.92) shows that W_{12} depends linearly on y, i.e.,

$$W_{12} = W_{121}(t) + W_{122}(t)y$$

Eq.(3.92) then becomes

$$-\frac{dW_2}{dt} + 2W_{122} = 0 \qquad (3.93)$$

We will make use of this equation later. The characteristic function now becomes:

$$W = W_{11}(t, y, u) + [W_{121}(t) + W_{122}(t)y]p_2$$
$$+ W_2(t)p_1$$

Putting into Eq.(3.90a) the new form of W and setting to zero terms with different powers of p_2, three equations are obtained:

$$p_2^0 \; : \; -\frac{\partial W_{11}}{\partial t} + \frac{\partial^2 W_{11}}{\partial y^2} = 0 \qquad (3.94a)$$

$$p_2^1 : \quad -\frac{dW_{121}}{dt} - \frac{dW_{122}}{dt}y + 2\frac{\partial^2 W_{11}}{\partial y \partial u} = 0 \tag{3.94b}$$

$$p_2^2 : \quad \frac{\partial^2 W_{11}}{\partial u^2} = 0 \tag{3.94c}$$

Eq.(3.94c) shows that

$$W_{11} = W_{111}(t, y) + W_{112}(t, y)u$$

Eq.(3.94b) then gives:

$$-\frac{dW_{121}}{dt} - \frac{dW_{122}}{dt}y + 2\frac{\partial W_{112}}{\partial y} = 0 \tag{3.95}$$

Therefore, W_{112} can be written as:

$$W_{112} = W_{1121}(t) + W_{1122}(t)y + W_{1123}(t)y^2$$

Eq.(3.95) becomes:

$$-\frac{dW_{121}}{dt} - \frac{dW_{122}}{dt}y + 2W_{1122} + 4yW_{1123} = 0 \tag{3.96}$$

Since all the W's in Eq.(3.96) are independent of y, we get:

$$-\frac{dW_{121}}{dt} + 2W_{1122} = 0 \tag{3.97a}$$

$$-\frac{dW_{122}}{dt} + 4W_{1123} = 0 \tag{3.97b}$$

Putting W_{11} into Eq.(3.94a), we get:

$$(-\frac{\partial W_{111}}{\partial t} + \frac{\partial^2 W_{111}}{\partial y^2}) - (\frac{dW_{1121}}{dt} - 2W_{1123})u$$

$$-\frac{dW_{1122}}{dt}yu - \frac{dW_{1123}}{dt}y^2 u = 0 \tag{3.98}$$

Since W_{111} is a function of t and y only and W_{1121}, W_{1122} and W_{1123} are functions of t only, the coefficients of u^0, u, yu, and $y^2\ u$ in Eq.(3.94b) should all be zero, which then gives

$$W_{1122} = c_1 \ ; \quad W_{1123} = c_2 \ ; \quad W_{1121} = 2c_2 t + c_3$$

and

$$\frac{\partial W_{111}}{\partial t} - \frac{\partial^2 W_{111}}{\partial y^2} = 0 \tag{3.99}$$

From Eqs.(3.97a) and (3.97b), we get:

$$W_{121} = 2c_1 t + c_4 \;;\quad W_{122} = 4c_2 t + c_5 \tag{3.100}$$

Also,from Eq.(3.93),

$$W_2 = 4c_2 t^2 + 2c_5 t + c_6$$

The final form of the characteristic function is therefore

$$\begin{aligned} W(t, y, u, p_1, p_2) = \; & W_{111}(t, y) \\ & +(2c_2 t + c_3 + c_1 y + c_2 y^2) u \\ & +[2c_1 t + c_4 + (4c_2 t + c_5) y] p_2 \\ & +(4c_2 t^2 + 2c_5 t + c_6) p_1 \end{aligned} \tag{3.101}$$

where $W_{111}(t, y)$ is any function satisfying the equation

$$\frac{\partial W_{111}}{\partial t} - \frac{\partial^2 W_{111}}{\partial y^2} = 0$$

We now conclude that the heat equation will be invariant under an infinitesimal group of transformations, if the characteristic function, W,is of the form, Eq.(3.101). Therefore, it should now be possible to reduce the number of independent variables by one.

Similarity transformations can be obtained by solving the subsystem

$$\frac{dt}{\xi} = \frac{dy}{\eta} = \frac{du}{\varsigma} \tag{3.102}$$

ξ, η and ς can be obtained from Eq.(3.83) and Eq.(3.101) as:

$$\begin{aligned} & \frac{dt}{4c_2 t^2 + 2c_5 t + c_6} \\ & = \frac{dy}{2c_1 t + c_4 + (4c_2 t + c_5) y} \\ & = \frac{du}{-W_{111}(t, y) - (2c_2 t + c_3 + c_1 y + c_2 y^2) u} \end{aligned} \tag{3.103}$$

We will consider a few special cases:

CASE 1:

$$W_{111}(t, y) = c_1 = c_2 = c_4 = c_6 = 0$$

$$(\mathit{Linear\ Group\ of\ Transformations})$$

Eq.(3.103) becomes

$$\frac{dt}{2c_5 t} = \frac{dy}{c_5 y} = \frac{du}{-c_3 u} \tag{3.104}$$

The two independent solutions to Eq.(3.104) are

$$\frac{y}{\sqrt{t}} = constant \;\; and \;\; \frac{u}{t^{\alpha}} = constant$$

where $\alpha = -c_3/(2c_5)$. The similarity transformation can now be written as

$$\eta = \frac{y}{\sqrt{t}} \;\; and \;\; f(\eta) = \frac{u}{t^{\alpha}} \tag{3.105}$$

The diffusion equation can be transformed into an ordinary differential equation

$$f'' = \alpha f - \frac{1}{2}\eta f' \tag{3.106}$$

These results are the same as those obtained by the Birkhoff-Morgan approach, Eq.(3.26), and the separation of variables method, Eq.(3.9). The deductive group approach systematically gives rise to a number of solutions, some of which cannot be discovered by inspectional group procedures.

CASE 2:

$$W_{111}(t, y) = c_1 = c_2 = c_4 = c_5 = 0$$

For this group, Eq.(3.103) becomes

$$\frac{dt}{c_6} = \frac{dy}{0} = \frac{du}{-c_3 u}$$

Following the same approach as in case 1, the similarity variables are found to be

$$\eta = y \;\; and \;\; f(\eta) = \frac{u}{exp(\beta t)} \tag{3.107}$$

where $\beta = -c_3/c_6$. The diffusion equation is transformed to:

$$f'' - \beta f = 0 \tag{3.108}$$

which is seen to be spiral group.

CASE 3:

$$W_{111}(t, y) = c_1 = c_3 = c_4 = c_5 = c_6 = 0$$

For this group, Eq.(3.103) becomes:

$$\frac{dt}{4t^2} = \frac{dy}{4ty} = \frac{du}{-(2t + y^2)u} \tag{3.109}$$

Now, the first two terms give

$$\frac{y}{t} = \frac{1}{k_1} \tag{3.110}$$

where k_1 is a constant. Next, substituting $t = k_1 y$ from Eq.(3.110) into the second and third terms in Eq.(3.109), we get:

$$-\frac{(2k_1 y + y^2)}{4k_1 y^2} dy = \frac{du}{u}$$

which, upon integration, gives

$$uy^{1/2} exp(\frac{y}{4k_1}) = constant$$

Using k_1 from Eq.(3.110), it becomes:

$$uy^{1/2} exp(\frac{y^2}{4t}) = k_2$$

The similarity variables and the diffusion equation are therefore:

$$\eta = \frac{y}{t} \quad and \quad f(\eta) = uy^{1/2} exp(\frac{y^2}{4t}) \tag{3.111}$$

$$\eta^2 f'' - \eta f' + \frac{3}{4} f = 0 \tag{3.112}$$

CASE 4:

$$W_{111}(t, y) = a_0 + a_1 t + \frac{1}{2} a_1 y^2$$

$$c_1 = c_2 = c_3 = c_4 = c_6 = 0$$

It can be shown that this special form of W_{111} satisfies Eq.(3.99). For this group, Eq.(3.103) becomes

$$\frac{dt}{2c_5 t} = \frac{dy}{c_5 y} = \frac{du}{-a_0 - a_1 t - (1/2) a_1 y^2} \tag{3.113}$$

from which

$$\eta = \frac{y}{\sqrt{t}} \quad ; \quad f(\eta) = u + \frac{a_1(2t + y^2)}{4c_5} - \frac{a_0}{2c_5} ln(t)$$

and the diffusion equation becomes:

$$f'' + \frac{1}{2}\eta f' - \frac{a_0}{2c_5} = 0 \tag{3.114}$$

More solutions to the heat equation are available in the works of Na and Hansen[7] and Bluman and Cole[6]. As pointed out earlier in this section, the underlying basis of the infinitesimal group method of Bluman-Cole and the characteristic function method of Na-Hansen is the same. The latter method expresses the infinitesimals or the transformation function in terms of a single function, W, which is the characteristic function. Also,since the finite group of transformations is generated by the infinitesimal group, we will mainly utilize the characteristic function method of Na and Hansen to find general similarity solutions wherever the deductive procedure is applied to partial differential models in the remainder of the book.

3.3 Summary

The methods for obtaining similarity transformation were classified into (a) direct methods and (b) group- theoretic methods. The direct methods such as separation of variables and dimensional analysis do not invoke group invariance. They are fairly straightforward and simple to apply. Group-theoretic methods on the other hand are mathematically more elegant, and the important concept of invariance under a group of transformations is always invoked. Again,in some group-theoretic procedures such as the Birkhoff-Morgan method and the Hellums-Churchill method the specific form of the group is assumed a priori. On the other hand,procedures such as the finite group method of Moran-Gaggioli, and the infinitesimal group methods of Bluman-Cole and Na-Hansen are deductive. In these procedures, a general group of transformations is defined and similarity solutions are systematically deduced.

REFERENCES

[1] Kline,S.J.,Similitude and Approximation Theory, McGraw-Hill (1965).

[2] Birkhoff,G.,Hydrodynamics,Princeton University Press (1960).

[3] Morgan,A.J.A.,"The Reduction by One of the Number of Independent Variables in Some Systems of Partial Differential Equations",Quar. J. Math.,Oxford,Vol.2,p.250 (1952).

[4] Hellums,J.D. and Churchill,S.W.,"Simplification of Mathematical Description of Boundary and Initial Value Problems",A.I.Ch.E. Journal,10 (1964).

[5] Moran.M.J. and Gaggioli,R.A.,A.I.A.A. Journal,Vol.6, p.2014(1968).

[6] Bluman,G.W. and Cole,J.D.,"General Similarity Solution of the Heat Equation",Journal of Mathematics and Mechanics,18,No.11 (1969).

[7] Na,T.Y. and Hansen,A.G.,"Similarity Analysis of Differential Equations by Lie Group", Journal of Franklin Institute,Vol.6 (1971).

[8] Sedov,L.I.,Similarity and Dimensional methods in Mechanics, Academic Press, Inc., New York (1959).

[9] Moran, M.J.,"A Generalization of Dimensional Analysis", Journal of Franklin Institute, Vol.6 (1971).

[10] Morrison,F.A.,Jr.,"General Dimensional and Similarity Analysis", Bul. Mech. Eng. Educ., Vol.8 (1969).

[11] Moran,M.J.,"A Unification of Dimensional Analysis via Group Theory", PhD Dissertation,University of Wisconsin,Madison,Wisconsin, 1967.

[12] Seshadri,R.,"Group Theoretic Method of Similarity Analysis Applied to Nonlinear Impact Probs", PhD Thesis, Univ. of Calgary,Alberta, Canada (1973).

[13] Moran,M.J.,Gaggioli,R.J. and Scholten,W.B.,"A Systematic Formalism for Similarity Analysis with Application to Boundary Layer Flow", Math. Res. Center, Rept.918, University of Wisconsin (1968).

[14] Moran,M.J. and Gaggioli,R.J.,"Similarity for a Real Gas Boundary layer Flow", Math. Res. Center, Rept. 919,University of Wisconsin (1968).

[15] Ames,W.F., Nonlinear Partial Dif. Eqs. in Engineering, ,Academic Press, New York (1972).

Chapter 4

APPLICATION OF SIMILARITY ANALYSIS TO PROBLEMS IN SCIENCE AND ENGINEERING

4.0 Introduction

Different methods for carrying out similarity analysis of partial differential equations were discussed in Chapter 3 with particular reference to the linear heat equation. The methods were classified into (1) direct methods and (2) group-theoretic methods. In the direct methods, the concept of group invariance is not explicitly invoked. They are straightforward and simple to apply. Since the direct methods are based on assumed transformations, the resulting solutions are restrictive. The group-theoretic methods on the other hand are based upon the invocation of invariance under groups of transformations of the partial differential equations and the auxiliary conditions. Group-theoretic methods such as Birkhoff-Morgan method and the Hellums-Churchill procedure start out by assuming a specific form of the group. Therefore, the resulting similarity solutions are restrictive. The simplicity of these methods is on account of the fact that only algebraic equations (resulting from invocation of invariance) need to be solved. On the other hand, deductive group procedures while being systematic and more rigorous and tedious.

In this chapter, we will apply these methods of similarity analysis to a variety of problems found in engineering science.

4.1 Laminar Two Dimensional Jet : Separation of Variables Method

The separation of variables method of Abbott and Kline[1] is applied to the problem of a steady two-dimensional incompressible laminar flow of fluid into an infinite region of the same fluid.

The equations of motion can be written as:

$$u\frac{\partial u}{\partial x} + v\frac{\partial u}{\partial y} = \nu\frac{\partial^2 u}{\partial y^2} \tag{4.1}$$

$$\frac{\partial u}{\partial x} + \frac{\partial v}{\partial y} = 0 \tag{4.2}$$

u and v are components of velocity in the x and y directions. ν is the kinematic viscosity.
The boundary conditions are:

$$y = 0 \ : \ v = 0 \ ; \ \frac{\partial u}{\partial y} = 0 \tag{4.3a}$$

$$y = \infty \ : \ u = 0 \tag{4.3b}$$

The total momentum flux across a section of the jet at any given value of x is constant.

$$2\rho \int_0^\infty u^2 \, dy = constant \tag{4.4}$$

ρ is the mass density of the fluid.

Introducing a stream function, ψ, such that

$$u = \frac{\partial \psi}{\partial y} \quad ; \quad v = -\frac{\partial \psi}{\partial x} \tag{4.5}$$

Eq.(4.2) is satisfied identically and Eq.(4.1) becomes

$$\frac{\partial \psi}{\partial y} \frac{\partial^2 \psi}{\partial x \partial y} - \frac{\partial \psi}{\partial x} \frac{\partial^2 \psi}{\partial y^2} = \nu \frac{\partial^3 \psi}{\partial y^3} \tag{4.6}$$

If we assume a transformation as follows:

$$\psi(x, y) = b\ g(x)\ f(\varsigma) \quad where\ \varsigma = \frac{ay}{\gamma(x)} \tag{4.7}$$

where a and b are constants; g(x) and $\gamma(x)$ are unknown functions, then Eq.(4.6) can be transformed into

$$\gamma g' - \gamma' g = \gamma g' \frac{ff''}{(f')^2} + \frac{\nu a}{b} \frac{f'''}{(f')^2} \tag{4.8}$$

The boundary conditions in the transformed coordinates then become:

$$\varsigma = 0 \ : \ bg' f(0) = 0$$

$$\varsigma = 0 \ : \ ba^2 \frac{g}{\gamma^2} f''(0) = 0 \tag{4.9}$$

$$\varsigma \to 0 \ : \ ba \frac{g}{\gamma} f'(\infty) = 0$$

For Eq.(4.8) to be reducible to an ordinary differential equation, we set

$$\gamma g' = constant = c_1$$

A second relation between γ and g can be found by using Eq.(4.4), which can be written as

$$\int_0^\infty u^2 \, dy = c_2 \tag{4.4a}$$

We have

$$u = \frac{\partial \psi}{\partial y} = ab \frac{g}{\gamma} f' \quad and \quad dy = \frac{\gamma}{a} d\varsigma$$

Therefore, Eq.(4.4a) becomes

$$b^2 a \frac{g^2}{\gamma} \int_0^{\infty} [f'(\varsigma)]^2 d\varsigma = c_2$$

or

$$\frac{g^2}{\gamma} = c_3 = constant$$

Solving for g and γ from the relationships

$$\gamma g' = c_1 \quad and \quad \frac{g^2}{\gamma} = c_2,$$

we have

$$g(x) = [3c_1 c_3 (x + x_0)]^{1/3} \tag{4.10a}$$

$$\gamma(x) = \frac{1}{c_3} [3c_1 c_3 (x + x_0)]^{2/3} \tag{4.10b}$$

where x_0 is a constant of integration.

It can now be seen that $\gamma' g$ in Eq.(4.8) is also a constant.

If c_1 and c_3 are chosen as

$$c_1 = \frac{\sqrt{\nu a}}{b} \quad and \quad c_3 = \frac{1}{3},$$

Eq.(4.8) can be rewritten as

$$f''' + ff'' + (f')^2 = 0 \tag{4.11}$$

with boundary conditions

$$f(0) = 0 \;\; ; \;\; f'(0) = 0 \;\; ; \;\; f'(\infty) = 0$$

and the similarity variable

$$\varsigma = \frac{ky}{3(x + x_0)^{2/3}}$$

where k is a constant. The solution to the above equations have been reported in closed form[2].

4.2 Impact of Rods With Nonlinear Material Properties: Separation of Variables Method

We now consider the problem of impact of thin long rods which exhibit nonlinear viscous and elastic behavior. The method of separation of variables is used to obtain the required similarity transformations for a system of partial differential equations[3]. The equations of motion for longitudinal deformation of a uniaxial rod are

$$\frac{\partial \sigma}{\partial x} = \rho \frac{\partial v}{\partial t} \tag{4.12a}$$

$$v = \frac{\partial u}{\partial t} \tag{4.12b}$$

$$e = \frac{\partial u}{\partial x} \tag{4.12c}$$

$$\frac{\partial e}{\partial t} = \frac{\partial}{\partial t}(\frac{\sigma}{\mu})^q + (\frac{\sigma}{\lambda})^n \tag{4.12d}$$

where σ is the normal stress along the axis of the rod; v is the component of the velocity along the axis;e is the strain;x is the space coordinate and t is the time. The material constants are q, n, μ and λ.

Eq.(4.12a) is the equilibrium equation,Eq.(4.12c) is the strain displacement relation, and Eq.(4.12d) is the constitutive relationship. When $\lambda \to \infty$,the stress-strain behavior for the nonlinear elastic material is obtained. If $\mu \to \infty$,however,the nonlinear viscous behavior would result. Materials that exhibit the relationship as described in Eq.(4.12d) are called Maxwell solids.

The stress, displacement and velocity is assumed to be as follows:

$$\begin{aligned} \sigma(x,t) &= S(t)F(\varsigma) \\ u(x,t) &= U(t)G(\varsigma) \\ v(x,t) &= V(t)H(\varsigma) \end{aligned} \tag{4.13}$$

where the similarity variable is defined as

$$\varsigma = \frac{x}{X(t)}$$

For nonlinear elastic materials, the governing differential equations are hyperbolic. For viscous materials, the equations are parabolic.

Substituting Eq.(4.13) into Eq.(4.12), we obtain the following:

$$F' = \delta H - \gamma \varsigma H' \tag{4.14a}$$

$$H = \beta G - \gamma \varsigma G' \tag{4.14b}$$

$$H' = \frac{q}{1+w}(\alpha F^q - \gamma \varsigma F^{q-1} F') + \frac{w}{1+w} F^n \tag{4.14c}$$

where

$$\alpha = \frac{US'}{VS} \;\; ; \;\; \beta = \frac{U'}{V} \;\; ;$$

$$\gamma = \frac{UX'}{VX} \;\; ; \;\; \delta = \frac{\rho XV'}{S} \;\; ;$$

$$\frac{1}{1+w} = \frac{X}{U}\left(\frac{S}{\mu}\right)^q \;\; ; \;\; \frac{w}{1+w} = \frac{X}{V}\left(\frac{S}{\lambda}\right)^n \tag{4.15}$$

Primes denote differentiation with respect to the argument.

If similarity solutions are to exist, then α, β,γ,δ and w must represent constant quantities. That is, time must not explicitly appear in the differential equations (4.14).

One set of solutions leading to problems of practical interest is given by

$$S(t) = At^{\alpha} \quad ; \quad U(t) = Bt^{\beta}$$

$$V(t) = Dt^{\delta} \quad ; \quad X(t) = Ct^{\gamma} \tag{4.16}$$

Substituting Eq.(4.16) into Eq.(4.12), the following relationships can be obtained:

$$\frac{\rho CD}{A} = 1 \quad ; \quad \gamma + \delta - \alpha = 1$$

$$B = D \quad ; \quad \beta - \delta = 1$$

$$\frac{C}{B}\left(\frac{A}{\mu}\right)^q = \frac{1}{1+w} \quad ; \quad \gamma - \delta + \alpha q = 1$$

$$\frac{C}{D}\left(\frac{A}{\lambda}\right)^n = \frac{w}{1+w} \quad ; \quad \gamma - \delta + \alpha n = 0 \tag{4.17}$$

The constants α,β,γ and δ can be determined for the constitutive model and the particular boundary conditions.

Two distinct types of boundary conditions are considered; the first is the application of velocity impact

$$v(0,t) = v_0\left(\frac{t}{t_0}\right)^{\delta} \tag{4.18a}$$

and the other is the application of stress impact

$$\sigma(0,t) = \sigma_0\left(\frac{t}{t_0}\right)^{\alpha} \tag{4.18b}$$

where σ_0, v_0 and t_0 are reference quantities.

For the Maxwell solid (λ and μ are finite), the solution for Eq.(4.17) is

$$\alpha = \frac{1}{q-n} \;\; ; \;\; \beta = 1 + \frac{1+q}{2(q-n)}$$

$$\gamma = 1 + \frac{1-q}{2(q-n)} \quad ; \quad \delta = \frac{1+q}{2(q-n)} \tag{4.19}$$

For the elastic material, $\lambda \to \infty$ which implies that $w \to 0$. For an applied velocity at the end of the elastic rod,

$$\alpha = \frac{2\delta}{1+q} \quad ; \quad \gamma = 1 + \left(\frac{1-q}{1+q}\right)\delta ,$$

where δ is defined in Eq.(4.18a).

For the applied stress problem, we express γ and δ in terms of α, as follows:

$$\gamma = 1 + \frac{\alpha}{2}(1-q) \quad ; \quad \delta = \frac{\alpha}{2}(1-q)$$

For the applied velocity problem,

$$v(0,t) = Dt^{\delta} H(0) = v_0 \left(\frac{t}{t_0}\right)^{\delta} \tag{4.20}$$

Therefore,

$$D = \frac{v_0}{t_0{}^{\delta}} \quad and \quad H(0) = 1$$

Using Eqs.(4.13),(4.16),(4.17) and (4.20),the following expressions can be obtained:

$$\sigma(x,t) = (\rho v_0{}^2 \mu^q)^{1/(1+q)} \left(\frac{t}{t_0}\right)^{2\delta/(1+q)} F(\varsigma) \tag{4.21}$$

Starting with the relationship $X(t) = Ct^{\gamma}$, C and γ can be determined. Then, the similarity variable ς can be written as

$$\varsigma = \frac{x}{X(t)}$$

where

$$X(t) = \left(\frac{\mu}{\rho}\right)^{q/(1+q)} v_0{}^{s} t_0 \left(\frac{t}{t_0}\right)^{1+\delta s}$$

with $s = (1-q)/(1+q)$.

Eqs.(4.14) uncouple and a single ordinary differential equation can be obtained as

$$\frac{1}{q}(G')^{(1-q)/q} G'' = \gamma^2 \varsigma^2 G''$$

$$+\gamma(\gamma + 1 - 2\beta)\varsigma G' + \beta(\beta - 1) G \tag{4.22}$$

At $\varsigma = 0, H(0) = 1$, which according to Eq.(4.14b) transforms to $G(0) = 1/(1+\delta)$.

At the wavefront, ς equals to ς_w, the continuity of the rod must be preserved. This implies zero displacement immediately ahead of the wavefront,i.e., $G(\varsigma_w) = 0$. A technique for determining the numerical value of ς_w is discussed in chapter 8.

For a finite impulse and displacement at the origin, both α and δ are greater than -1. For a linear elastic material, q=1, $\varsigma_w = 1$.

$$G(\varsigma) = \frac{1}{1+\delta}(1-\varsigma)^{1+\delta} \tag{4.23}$$

A similar procedure can be followed for the stress problem. Other solutions to the impact problem are given in references (3) and (4).

4.3 Diffusion of Vorticity From a Line Vortex Immersed in a Quiescent Fluid : Dimensional Method

The equation for the diffusion of a vorticity in a fluid is given by:

$$\nu\left(\frac{\partial^2 \omega}{\partial r^2} + \frac{1}{r}\frac{\partial \omega}{\partial r}\right) = \frac{\partial \omega}{\partial t} \tag{4.24}$$

where ω is the vorticity, r is the radius about the origin, ν is the kinematic viscosity and t is time. The circulation of flow,Γ_0, is assumed to be constant as the radius of the circular cylinder of radius ,r_0, rotating in the fluid approaches zero.

$$\Gamma_0 = \int_0^r \int_0^{2\pi} \omega r dr d\theta \tag{4.25}$$

The line vortex is assumed to be suddenly introduced into a fluid at rest. At large times, the entire body of the fluid would rotate as a vortex with a circumferential velocity

$$v_\theta = \frac{\Gamma_0}{2\pi r} \tag{4.26}$$

Eq.(4.25) can now be rewritten by letting $\Omega = \omega/\Gamma_0$ as follows:

$$\int_0^r \int_0^{2\pi} \Omega r dr d\theta = 1 \tag{4.27}$$

The vorticity of the fluid at rest is zero, and the vorticity at any point (excluding the origin) would vanish at large times. The problem of interest here is to determine the variation of vorticity and flow velocity at any time t.

From Eqs.(4.24) and (4.25) it is clear that

$$\omega = f(r, t, \Gamma_0, \nu)$$

or

$$\Omega = F(r, t, \nu) \tag{4.28}$$

The dimensional matrix for the relationship

$$\Omega^{a_1} r^{a_2} t^{a_3} \nu^{a_4} = M^0 L_r{}^0 T^0$$

can be written as follows:

	a_1	a_2	a_3	a_4
	Ω	r	t	ν
M	0	0	0	0
L_r	-2	1	0	2
T	0	0	1	1

The number of variables is 4, and the rank of the dimensional matrix is equal to 2. Therefore, the two Pi terms can be written as

$$\pi_1 = \frac{r}{\sqrt{\nu t}} \quad ; \quad \pi_2 = \Omega \nu t$$

The similarity transformation can be expressed as:

$$\pi_1 = \phi(\pi_2{}^2)$$

or

$$\Omega(r,t) = \frac{1}{\nu t}[\phi(\varsigma)] \tag{4.29}$$

where $\varsigma = r^2/(\nu t)$.

The resulting ordinary differential equation can be obtained by substituting Eq.(4.29) into Eq.(4.24) using the relationship, $\omega = \Gamma_0 \Omega$, as

$$4\varsigma\phi'' + 4\phi' + (\phi + \varsigma\phi') = 0 \tag{4.30}$$

Integrating once,

$$4\varsigma\phi' + \varsigma\phi = c_1$$

Since ϕ and ϕ' are finite for any given time, and therefore any ς, then as $\varsigma \rightarrow 0, c_1 = 0$. Integrating again,

$$\phi(\varsigma) = c_2 exp(-\varsigma^2/4) \tag{4.31}$$

Using the boundary condition, Eq.(4.27),

$$c_2 = \frac{1}{4\pi[1 - exp(-r^2/4\nu t)]}$$

When $r \rightarrow \infty, c_2 = 1/(4\pi)$, therefore the required similarity solution is given by

$$\omega(r,t) = \frac{\Gamma_0}{4\pi\nu t} exp(-r^2/4\nu t) \tag{4.32}$$

4.4 Laminar Boundary Layer Equation : Dimensional Method

Consider the problem of an incompressible boundary layer flow, on a flat plate with a uniform free stream U_0. The flow is governed by the continuity equation

$$\frac{\partial u}{\partial x} + \frac{\partial v}{\partial y} = 0 \tag{4.33a}$$

and the momentum equation

$$u\frac{\partial u}{\partial x} + v\frac{\partial u}{\partial y} = \nu\frac{\partial^2 u}{\partial y^2} \tag{4.33b}$$

The boundary conditions can be written as:

$$y = 0 \ : \ u = v = 0 \quad ; \quad y \rightarrow \infty \ : \ u \rightarrow U_0$$

We can assume the dimensional relationship as

$$u = u(x, y, \nu, U_0) \quad ; \quad v = v(x, y, \nu, U_0) \tag{4.34}$$

Traditional dimensional analysis in which no distinction is made between the x and y directions of the length gives the following relationship :

$$\frac{u}{U_0} = f_1\left(\frac{y}{x}, \frac{U_0 x}{\nu}\right) \tag{4.35}$$

Eq.(4.35) is not a similarity transformation. However, assigning separate dimensions to the x and y length directions, the dimensional matrix for the relationship

$$u^{a_1} x^{a_2} y^{a_3} \nu^{a_4} U_0{}^{a_5} = M^0 L_x{}^0 L_y{}^0 T^0$$

can be written as

	a_1	a_2	a_3	a_4	a_5
	u	x	y	ν	U_0
M	0	0	0	0	0
L_x	1	1	0	0	1
L_y	0	0	1	2	0
T	-1	0	0	-1	-1

where the equal sign means "dimensionally equal to". The rank and the number of pi terms in the dimensional matrix are 3 and 2, respectively.

The Pi terms are

$$\pi_1 = \frac{u}{U_0} \quad ; \quad \pi_2 = y\left(\frac{U_0}{\nu x}\right)^{1/2}$$

Therefore, the similarity transformation is

$$\frac{u}{U_0} = f_2\left[y\left(\frac{U_0}{\nu x}\right)^{1/2}\right]$$

Similarly,

$$v\left(\frac{x}{\nu U_0}\right)^{1/2} = g_2\left[y\left(\frac{U_0}{\nu x}\right)^{1/2}\right] \tag{4.36}$$

4.5 Free Convection from a Vertical Needle:Hellums and Churchill Method

Consider a vertical solid axisymmetrical body of radius,$R(x)$,whose surface temperature or local heat flux may vary vertically as shown in Fig.4.1. It is assumed in the formulation of this problem that the temperature increase is slight. Furthermore, the Prandtl's boundary layer approximation is made. The body is also assumed to be thin so that the square of its slope $[R'(x)^2]$ is negligible. This last approximation facilitates a wide class of self-similar solutions.

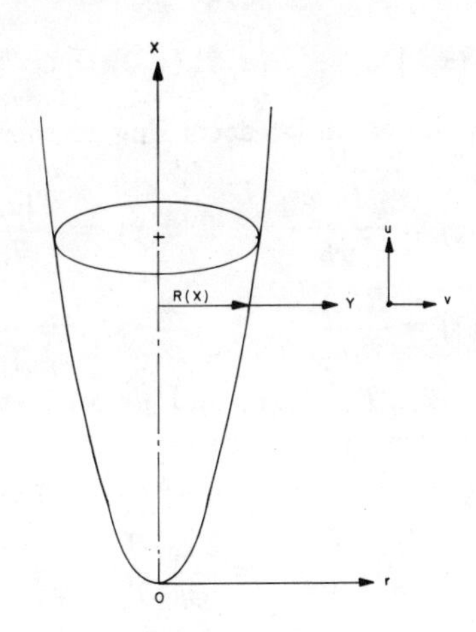

Fig.4.1 Free Convection From a Thin Vertical Needle

The governing differential equations are[5]

$$[(R+y)u]_x + [(R+y)v]_y = 0$$

$$uu_x + vu_y = \frac{\nu}{R+y}[(R+y)u_y]_y + g\beta T \tag{4.37}$$

$$uT_x + vT_y = \frac{\alpha}{R+y}[(R+y)T_y]_y$$

where ν is the kinematic viscosity, α is the thermal diffusivity, and β is the coefficient of thermal expansion. The boundary conditions at the surface and at infinity are:

$$u(x,0) = v(x,0) = 0$$

$$T(x,0) \quad or \quad -R(x)T_y(x,0) \quad is \ prescribed$$

$$u(x,\infty) = T(x,\infty) = 0 \tag{4.38}$$

Introducing the Stokes stream function, ψ, such that

$$(R+y)u = \psi_y \quad ; \quad (R+y)v = -\psi_x$$

The first of Eq.(4.37) is satisfied. The remaining two equations can be written as

$$(\psi_y \frac{\partial}{\partial x} - \psi_x \frac{\partial}{\partial y}) \frac{\psi_y}{R+y} = \nu \frac{\partial}{\partial y}[(R+y)\frac{\partial}{\partial y}(\frac{\psi_y}{(R+y)})]$$

$$+ g\beta(R+y)T$$

$$(\psi_y \frac{\partial}{\partial x} - \psi_x \frac{\partial}{\partial y})T = \alpha \frac{\partial}{\partial y}((R+y)T_y) \tag{4.39}$$

The boundary conditions can be written as:

$$\psi(x,0) = \psi_y(x,0) = \lim_{y \to \infty} \frac{\psi_y}{y} = T(x,\infty) = 0$$

$$T(x,0) \quad or \quad -R(x)T_y(x,0) = Qx^m \tag{4.40}$$

We now introduce the variables according to the affine transformation as follows:

$$\bar{\psi}(\bar{x},\bar{y}) = \frac{\psi(x,y)}{\psi_0} \quad ; \quad \bar{T}(\bar{x},\bar{y}) = \frac{T(x,y)}{T_0}$$

$$\bar{R}(\bar{x}) = \frac{R(x)}{R_0} \quad ; \quad \bar{x} = \frac{x}{x_0} \quad ; \quad \bar{y} = \frac{y}{y_0} \tag{4.41}$$

The reference variables ψ_0, T_0, R_0, x_0 and y_0 are determined such that a minimum parametric description results.

Transforming Eq.(4.39) by using Eq.(4.41), we get

$$(\bar{\psi}_{\bar{y}} \frac{\partial}{\partial \bar{x}} - \bar{\psi}_{\bar{x}} \frac{\partial}{\partial \bar{y}})[\frac{\bar{\psi}_{\bar{y}}}{S(\bar{y})}] = \frac{\nu x_0}{\psi_0} \frac{\partial}{\partial \bar{y}}(S(\bar{y}) \frac{\partial}{\partial \bar{y}}(\frac{\bar{\psi}_{\bar{y}}}{S(\bar{y})})$$

$$+ \frac{g\beta_0 T_0 x_0 y_0^4}{\psi_0^2}[(\frac{R_0}{y_0})\bar{R} + \bar{y}]\bar{T}$$

$$(\bar{\psi}_{\bar{y}} \frac{\partial}{\partial \bar{x}} - \bar{\psi}_{\bar{x}} \frac{\partial}{\partial \bar{y}})\bar{T} = \frac{\alpha x_0}{\psi_0}[(\frac{R_0}{y_0})\bar{R} + \bar{y}]\bar{T}_{\bar{y}}$$

The boundary conditions are transformed into:

$$\bar{\psi}(\bar{x},0) = \bar{\psi}_{\bar{y}}(\bar{x},0) \lim_{\bar{y}\to 0}(\frac{\bar{\psi}_{\bar{y}}}{\bar{y}} = \bar{T}(\bar{x},\infty) = 0$$

$$\bar{T}(\bar{x},0) = \frac{Qx_0^m}{T_0}(\bar{x}^m)$$

The parametric form of the solution can be expressed as:

$$\psi = \psi_0 f(\frac{x}{x_0}; \frac{y}{y_0} | \frac{R_0}{y_0}; \frac{\nu x_0}{\psi_0}; \frac{g\beta T_0 x_0 y_0^4}{\psi_0^2}, \frac{\alpha x_0}{\psi_0} | \frac{Qx_0^m}{T_0})$$

$$T = T_0 g(\frac{x}{x_0}; \frac{y}{y_0} | \frac{R_0}{y_0}; \frac{\nu x_0}{\psi_0}; \frac{g\beta T_0 x_0 y_0^4}{\psi_0^2}, \frac{\alpha x_0}{\psi_0} | \frac{Qx_0^m}{T_0}) \tag{4.42}$$

For minimum parametric description, we set

$$\frac{Qx_0^m}{T_0} = 1 \ ; \ \frac{g\beta T_0 x_0 y_0^4 = 1}{\psi_0^2} \ ; \ \frac{\nu x_0}{\psi_0} = 1 \ ; \ \frac{R_0}{y_0} = 1$$

Recognizing that $Pr=\nu/\alpha$ is the Prandtl's number, the similarity transformation is obtained by suitably eliminating the reference variables:

$$\psi(x,y) = \nu x f(\varsigma; Pr)$$

$$T(x,y) = Qx^m g(\varsigma; Pr)$$

where

$$\varsigma = \frac{y}{(\frac{\nu^2}{g\beta Q})^{1/4} x^{(1-m)/4}} \tag{4.43}$$

Also,the following relationship holds for the variations of $R(x)$:

$$R(x) = K(\frac{\nu^2}{g\beta Q})^{1/4} x^{\frac{1-m}{4}} \tag{4.44}$$

where K is a dimensionless parameter related to the Grashof number.

Substituting Eq.(4.43) into Eqs.(4.39) and (4.40), the following ordinary differential equations can be obtained:

$$f''' - (1-f)(\frac{f'}{K+\varsigma})' - \frac{1+m}{2}\frac{(f')^2}{K+\varsigma} + (K+\varsigma)g = 0 \tag{4.45a}$$

$$[(K+\varsigma)g']' + Pr(fg' - mf'g) = 0 \tag{4.45}$$

with the transformed boundary conditions as:

$$f(0) = f'(0) = \lim_{\varsigma\to\infty}[\frac{f'(\varsigma)}{\varsigma}] = g(\infty) = 0 \tag{4.46a}$$

$$g(0) \quad or \quad -Kg'(0) = 1 \tag{4.46b}$$

Analytical solutions for Eqs.(4.45) and (4.46) were discovered by Yih[6] for the case when m=-1 and K=0 (with g(0)=1). For Pr=1, his solutions can be written as:

$$f(\varsigma) = \frac{6\varsigma^2}{\varsigma^2 + \sqrt{96}} \quad ; \quad g(\varsigma) = \frac{1}{(1 + \varsigma^2/\sqrt{96})^3}$$

and for Pr=2,

$$f(\varsigma) = \frac{\varsigma^2}{2(1 + \varsigma^2/8)} \quad ; \quad g(\varsigma) = \frac{1}{(1 + \varsigma^2/8)^4}$$

For the general problem, numerical integration may be used to obtain the solution.

Since the Hellums-Churchill method utilizes an affine transformation, the method could yield new and significant results even in the relatively well-trodden fields of fluid mechanics and heat transfer. Solutions available by this method remain hidden from the traditional dimensional procedures (MLT system of dimensions). The familiar Falkner Skan family of solutions of boundary layer equations is an example.

4.6 Deflection of a Semi-Infinite Wedge Shaped Plate : Birkhoff-Morgan Method

The one parameter linear group of transformations is used in this example to obtain the transverse deflection of a semi- infinite wedge-shaped plate of constant thickness, which is clamped along one edge and is free at the other, as shown in Fig.4.2.

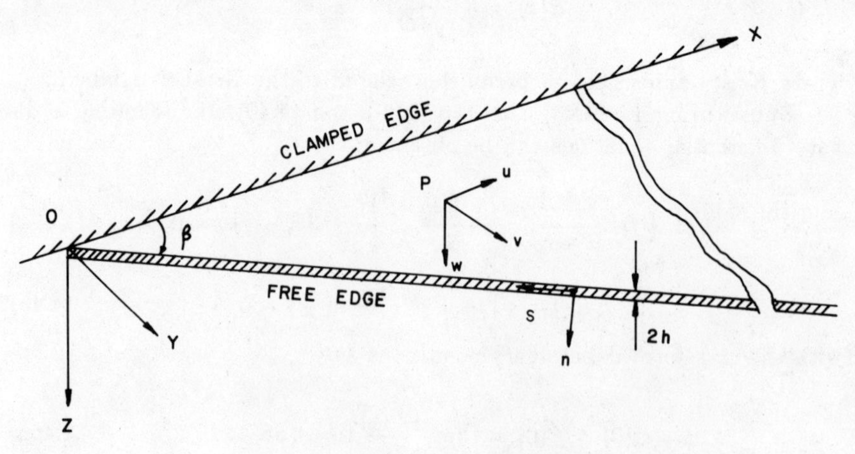

Fig.4.2 Wedge Plate Bending

The plate equation for small deflections can be expressed as:

$$\nabla^4 w \equiv \frac{\partial^4 w}{\partial x^4} + 2\frac{\partial^4 w}{\partial x^2 \partial y^2} + \frac{\partial^4 w}{\partial y^4}$$

$$= \frac{q(x,y)}{D} \tag{4.47}$$

where $w(x,y)$ is the deflection of the plate, D is the bending rigidity, and $q(x,y)$ is the transverse load per unit area.

We now define a one-parameter linear group G as follows:

$$\bar{w} = A^p w \ ; \ \bar{x} = A^m x \ ; \ \bar{t} = A^n t \tag{4.48}$$

where A is the parameter; m, n and p are constants to be determined by invoking invariance of Eq.(4.47) under G , such that

$$\nabla^4 \bar{w} - \frac{q(\bar{x},\bar{y})}{D} = \nabla^4 w - \frac{q(x,y)}{D} \tag{4.49}$$

Therefore,

$$A^{p-4m}\left(\frac{\partial^4 w}{\partial x^4} + A^{p-2m-2n}\left(2\frac{\partial^4 w}{\partial x^2 \partial y^2}\right)\right.$$

$$\left.+A^{p-2n}\left(\frac{\partial^4 w}{\partial y^4}\right) = \frac{q(x,y)}{D} \tag{4.50}$$

For invariance,

$$\frac{m}{n} = 1 \ \ and \ \ \frac{p}{m} = 4$$

It is also noted that $q(x,y) = q(\bar{x},\bar{y}) = q_0$ (a uniform plate loading). The absolute invariants are obtained by determining r and s such that:

$$\bar{x}\bar{t}^r = xt^r \ ; \ \bar{w}\bar{x}^s = wx^s \tag{4.51}$$

Therefore,

$$r = -\frac{m}{n} = -1 \ ; \ s = -\frac{p}{m} = -4$$

and the similarity transformation can be expressed as

$$w(x,y) = x^4 f(\varsigma) \ \ where \ \varsigma = \frac{y}{x} \tag{4.52}$$

Substitution of Eq.(4.52) into Eq.(4.47) gives the following ordinary differential equation:

$$(\varsigma^2+1)^2\frac{d^4 f}{d\varsigma^4} - 4\varsigma(\varsigma^2+1)\frac{d^3 f}{d\varsigma^3} + 4(3\varsigma^2+1)\frac{d^2 f}{d\varsigma^2}$$

$$-24\varsigma\frac{df}{d\varsigma}+24f=\frac{q_0}{D} \tag{4.53}$$

The general solution for Eq.(4.53) has been found as [7]:

$$f(\varsigma)=C_1\varsigma+C_2(\varsigma^2-\frac{1}{3})+C_3\varsigma^3+C_4(\varsigma^4-3\varsigma^2)+\frac{q_0\varsigma^2}{8D} \tag{4.54}$$

For the clamped edge, the slope and deflection is zero, therefore, w=0 and $w_y=0$. In terms of similarity variables,

$$f(0)=f'(0)=0 \tag{4.55a, b}$$

For the free edge at an angle of β to the x axis, the bending moment and shear force are zero. Therefore, along $\varsigma=tan(\beta)$, the bending moment can be expressed as

$$\nu\nabla^2 w\ sin(\beta)+(1-\nu)\Big[\frac{\partial^2 w}{\partial x^2}cos^2\beta+\frac{\partial^2 w}{\partial y^2}sin^2\beta$$

$$-\frac{\partial^2 w}{\partial x\partial y}sin(2\beta)\Big]=0 \tag{4.56c}$$

and the vanishing shear force as

$$-\frac{\partial}{\partial x}\nabla^2 w\ sin\beta+\frac{\partial}{\partial y}\nabla^2 w\ cos\beta-(1-\nu)\frac{\partial}{\partial s}$$

$$\Big[\frac{\partial^2 w}{\partial x\partial y}cos(2\beta)+\frac{1}{2}(\frac{\partial^2 w}{\partial y^2}-\frac{\partial^2 w}{\partial x^2})\Big]=0 \tag{4.56d}$$

Using conditions $f(0)=f'(0)=0$, both C_1 and C_2 are found to be zero. Therefore,

$$f(\varsigma)=C_3\varsigma^3+C_4(\varsigma^4-3\varsigma^2)+\frac{q_0\varsigma^2}{8D} \tag{4.57}$$

If $y/x=\tan(\beta)=k$ (constant), then along ς=k Eqs.(4.56c) and (4.56d) become:

$$12(1+\nu k^2)f(k)-6k[\nu k+(2-\nu)](\frac{df}{d\varsigma})_{\varsigma=k}$$

$$+[\nu k^4+2(2-\nu)k^2+\nu](\frac{d^2f}{d\varsigma^2})_{\varsigma=k}=0 \tag{4.58a}$$

and

$$24k(1+\frac{1-\nu}{1+k^2})f(k)+6[3k^2+(2-\nu)(\frac{df}{d\varsigma})_{\varsigma=k}$$

$$-6(1+k^2)(\frac{d^2f}{\partial\varsigma^2})_{\varsigma=k}+(1+k^2)^2(\frac{d^3f}{d\varsigma^3})_{\varsigma=k}=0 \tag{4.58b}$$

Substituting $f(\varsigma)$ from Eq.(4.57) into Eqs.(4.58a) and (4.58b), we get two simultaneous equations:

$$[8k^6 + 7k^4(2-\nu) + 3k^2(1-\nu) + 1]C_3 + 2k[k^2(5-3\nu)$$

$$+(1+\nu)]\,C_4 = -\frac{kP}{4D}[4k^4 + k^2(5-\nu)$$

$$+2k(1-\nu) + (1-\nu)]$$

and

$$6\nu k(k^2+1)C_3 + 2[k^4(6-\nu) + 2k^2(1+\nu) - \nu]C_4$$

$$= \frac{P}{4D}[2k^2(1-2\nu) - \nu k^4 - \nu]$$

Therefore, constants C_3 and C_4 can be obtained.

The deflection, w(x,y), can be written as

$$w(x,y) = x^4\left[C_3\left(\frac{y}{x}\right)^3 + C_4\left(\frac{y}{x}\right)^4 - 3C_4\left(\frac{y}{x}\right)^2 + \frac{q_0}{8D}\left(\frac{y}{x}\right)^2\right] \tag{4.59}$$

Other boundary conditions for the edges of the plate can be similarly considered.

4.7 Heated Jet: Birkhoff-Morgan Method[8]

A laminar jet of an incompressible fluid emerging from a narrow slot or a circular hole and mixing with the surrounding fluid(see Fig.4.3), where the physical properties are temperature dependent, have applications in liquid metal injection and high temperature arcs.

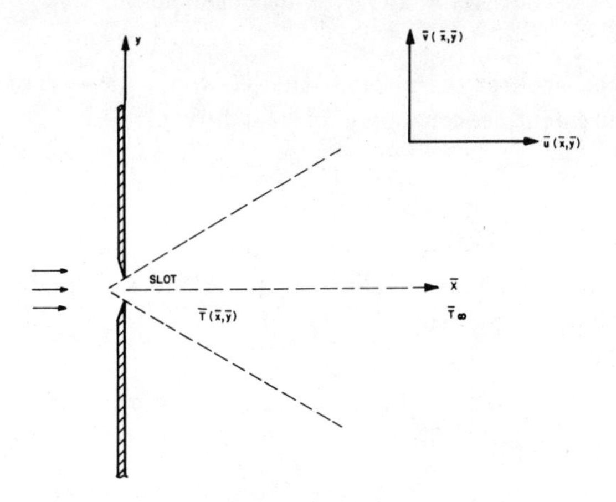

Fig.4.3 Heated Jet

The two-dimensional laminar jet flow of the incompressible fluid in a medium at rest with infinite plane boundary and at constant temperature is considered. The dependence of viscosity and thermal conductivity on temperature for constant pressure and steady flow is assumed to be

$$\mu = \mu_0 f(\theta) \quad ; \quad k = k_0 \phi(\theta) \tag{4.60}$$

The equation of motion can be written as

$$\frac{\partial \bar{u}}{\partial \bar{x}} + \frac{\partial \bar{v}}{\partial \bar{y}} = 0 \tag{4.61a}$$

$$\rho\left(\bar{u}\frac{\partial \bar{u}}{\partial \bar{x}} + \bar{v}\frac{\partial \bar{u}}{\partial \bar{y}}\right) = \frac{\partial}{\partial \bar{y}}\left(\mu(\bar{\theta})\frac{\partial \bar{u}}{\partial \bar{y}}\right) \tag{4.61b}$$

$$\rho C_p\left(\bar{u}\frac{\partial \bar{\theta}}{\partial \bar{x}} + \bar{v}\frac{\partial \bar{\theta}}{\partial \bar{y}}\right) = \frac{\partial}{\partial \bar{y}}\left[k(\bar{\theta})\frac{\partial \bar{\theta}}{\partial \bar{y}} + \left(\frac{\partial \bar{u}}{\partial \bar{y}}\right)^2\right] \tag{4.61c}$$

$\bar{u}$ and $\bar{v}$ are components of velocity in the $\bar{x}$ and $\bar{y}$ directions,ρ is mass density,$\bar{\theta}$ is the temperature,$\mu(\bar{\theta})$ is the viscosity and $k(\bar{\theta})$ is the thermal conductivity.
The boundary conditions are:

$$\bar{y} = 0 : \bar{v} = \frac{\partial \bar{u}}{\partial \bar{y}} = \frac{\partial \bar{\theta}}{\partial y} = 0 \tag{4.62a}$$

$$\bar{y} = \infty : \bar{u} = \bar{\theta} = 0 \tag{4.62b}$$

The momentum flux across any cross-section perpendicular to the jet axis will be assumed to be constant,i.e.,

$$M_0 = 2\rho \int_0^\infty \bar{u}^2 \, d\bar{y} = constant \tag{4.63}$$

Had there been no heat dissipation, the jet would have satisfied one more physical requirement,i.e.,constancy of heat flux:

$$Q = \int_0^\infty \rho C_p \bar{\theta} \bar{u} d\bar{y} = constant \tag{4.64}$$

We will now nondimensionalize the equations and conditions by introducing the following quantities:

$$\alpha = \frac{M_0}{2\rho\nu_0^2} \quad ; \quad x = \alpha\bar{x} \quad ; \quad y = \alpha\bar{y}$$

$$u = \frac{\bar{u}}{\alpha\nu_0} \quad ; \quad v = \frac{\bar{v}}{\alpha\nu_0} \quad ; \quad \lambda = \frac{\rho\nu_0 C_p}{k_0} \quad ; \quad D = \frac{Q}{\rho\alpha^2\nu_0^3} \tag{4.65}$$

where $\nu_0 = \mu_0/\rho$ and C_p is the specific heat at constant pressure.

We now introduce stream function,ψ, such that

$$u = \frac{\partial \psi}{\partial y} \quad ; \quad v = -\frac{\partial \psi}{\partial x} \tag{4.66}$$

Therefore, Eq.(4.61a) is satisfied and Eqs.(4.61a),(4.61b),(4.63) and (4.64) can be written as:

$$\frac{\partial \psi}{\partial y}\frac{\partial^2 \psi}{\partial x \partial y} - \frac{\partial \psi}{\partial x}\frac{\partial^2 \psi}{\partial y^2} = \frac{\partial}{\partial y}\left[f(\theta)\frac{\partial^2 \psi}{\partial y^2}\right] \tag{4.67a}$$

$$\frac{\partial \psi}{\partial y}\frac{\partial \theta}{\partial x} - \frac{\partial \psi}{\partial x}\frac{\partial \theta}{\partial y} = \frac{1}{\lambda}\frac{\partial}{\partial y}\left[\phi(\theta)\frac{\partial \theta}{\partial y}\right]$$

$$+f(\theta)\left(\frac{\partial^2 \psi}{\partial y^2}\right)^2 \tag{4.67b}$$

$$\int_0^\infty \left(\frac{\partial \psi}{\partial y}\right)^2 dy = 1 \tag{4.67c}$$

$$\int_0^\infty \left(\frac{\partial \psi}{\partial y}\right)\theta \, dy = 0 \tag{4.67d}$$

We will now invoke invariance of Eqs.(4.67) under a one-parameter linear group of transformations:

$$X = A^{\alpha_1} x \quad ; \quad Y = A^{\alpha_2} y$$

$$\Psi = A^{\beta_1} \psi \quad ; \quad \Theta = A^{\beta_2} \theta$$

α_1,α_2,β_1 and β_2 will be determined by imposing conformal invariance on the equations.

Since linear groups have been assumed at the outset, we stipulate that

$$f(\theta) = \theta^\gamma \;\; and \;\; \phi(\theta) = \theta^\delta$$

The conformal invariance relationships are:

$$\alpha_1 + 2\alpha_2 - 2\beta_1 = 3\alpha_2 - \beta_1 - \beta_2\gamma$$

$$\alpha_1 + \alpha_2 - \beta_1 - \beta_2 = 2\alpha_2 - \beta_2\delta - \beta_2$$

$$= -\beta_2\gamma + 2(2\alpha_2 - \beta_1)$$

$$\alpha_2 - 2\beta_1 = 0 \quad ; \quad \beta_1 + \beta_2 = 0 \tag{4.68}$$

Case (i):No Viscous Heat Dissipation

Eq.(4.68) can be solved, and the invariance can be determined to give the following similarity transformations:

$$\psi = x^{\frac{1}{3+\gamma}} F(\varsigma) \quad ; \quad \theta = x^{-\frac{1}{3+\gamma}} G(\varsigma)$$

where the similarity variable

$$\varsigma = \frac{y}{x^{2/(3+\gamma)}} \tag{4.69}$$

The transformed ordinary differential equations are

$$(F')^2 + FF'' = -(3+\gamma)\frac{d}{d\varsigma}(G^{\gamma} F'') \tag{4.70a}$$

$$F' G + F G' = -\frac{3+\gamma}{\lambda}\frac{d}{d\varsigma}(G^{\gamma} G') \tag{4.70b}$$

$$\int_0^{\infty} (F')^2 d\varsigma = 1 \tag{4.70c}$$

$$\int_0^{\infty} (F' G) d\varsigma \tag{4.70d}$$

The other boundary conditions are:

$$\varsigma = 0 \ : \ F = F'' = G' = 0 \tag{4.71a}$$

$$\varsigma = \infty \ : \ F' = G = 0 \tag{4.71b}$$

Case (ii) Viscous Dissipation Present

When viscous dissipation is present, we ignore Eq.(4.64) and solve the remainder of the equations and boundary conditions. Invocation of conformal invariance would lead to the following relationships:

$$\alpha_1 - \alpha_2 - \beta_1 + \beta_2\gamma = 0$$

$$\alpha_1 - \alpha_2 - \beta_1 + \beta_2\delta = 0$$

$$2\alpha_2 - 2\beta_1 + \beta_2(1 - \gamma + \delta) = 0$$

$$\alpha_2 - 2\beta_1 = 0 \tag{4.72}$$

The similarity transformation for $\beta_2 \neq 0, \gamma = \delta$ is given by

$$\psi = x^{\frac{1}{3+2\gamma}} F(\xi) \quad ; \quad \theta = x^{-\frac{2}{3+2\gamma}} G(\xi)$$

$$\xi = \frac{y}{x^{\frac{2}{3+2\gamma}}} \tag{4.73}$$

The ordinary differential equations for this case can be written as:

$$(F')^2 + FF'' = -(3+2\gamma)(G^{\gamma}F'')'$$

$$2F'G - FG' = -\frac{3+2\gamma}{\lambda}[(G^{\gamma}G')' + \gamma G^{\gamma}F''] \tag{4.74b}$$

$$\int_0^{\infty} (F')^2 d\xi = 1 \tag{4.74c}$$

with the boundary conditions:

$$\xi = 0 \; : \; F = F' = G' = 0 \tag{4.75a}$$

$$\xi = \infty \; : \; F' = G = 0 \tag{4.75b}$$

Consider the spiral group of transformations

$$\bar{x} = x + \alpha_1 a \;\; ; \;\; \bar{y} = e^{\alpha_2 a} y$$

$$\bar{\psi} = e^{\beta_1 a}\psi \;\; ; \;\; \bar{\theta} = e^{\beta_2}\theta$$

Conformal invariance of Eqs.(4.67a),(4.67b) and (4.67c) would give rise to the following relationships:

$$2\alpha_2 - 2\beta_1 = 3\alpha_2 - \beta_2\gamma - \beta_1$$

$$\alpha_2 - \beta_1 - \beta_2 = 2\alpha_2 - \beta_2\delta - \beta_2$$

$$= -\beta_2\gamma + 2(2\alpha_2 - \beta_1)$$

$$\alpha_2 - 2\beta_1 = 0 \;\; ; \;\; \beta_1 + \beta_2 = 0$$

When there is no dissipation condition, similarity exists only for $\gamma = \delta = -3$. Letting $p = \alpha_2/\alpha_1$, where p is some constant, and

$$\beta_1 = \frac{p\alpha_1}{2} \;\; and \;\; \beta_2 = -\frac{p\alpha_1}{2}$$

The similarity transformation can be written as:

$$\eta = \frac{y}{e^{px}} \;\; ; \;\; \psi = e^{px/2}F(\eta) \;\; ; \;\; \theta = e^{-px/2}G(\eta)$$

It can be verified that similarity solutions do not exist under linear or the spiral group for the dependence $f = e^{\gamma\theta}$ *and* $\phi = e^{\delta\theta}$.

4.8 Unsteady One-Dimensional Gas Dynamics Equations: Characteristic Function Method[9]

Consider the unsteady one-dimensional flow of ideal gas with constant entropy. The basic equations can be written as

$$\frac{\partial u}{\partial t} + u\frac{\partial u}{\partial x} + \frac{2}{k-1}a\frac{\partial a}{\partial x} = 0$$

$$\frac{\partial a}{\partial t} + \frac{k-1}{2}a\frac{\partial u}{\partial x} + u\frac{\partial a}{\partial x} = 0 \tag{4.76b}$$

This system of differential equations can be considered as two algebraic equations with eight variables, namely, u,a,x,t, p_1, p_2, q_1 and q_2, where

$$p_1 = \frac{\partial u}{\partial x}\ ;\ p_2 = \frac{\partial a}{\partial x}\ ;\ q_1 = \frac{\partial u}{\partial t}\ \textit{and}\ q_2 = \frac{\partial a}{\partial t}$$

Thus, Eqs.(4.76) become:

$$G_1 \equiv q_1 + up_1 + \frac{2}{k-1}ap_2 = 0 \tag{4.77a}$$

$$G_2 \equiv q_2 + \frac{k-1}{2}ap_1 + up_2 = 0 \tag{4.77b}$$

We now define an infinitesimal group of transformations as follows:

$$\bar{x} = x + \epsilon\alpha(u,a,x,t) + O(\epsilon^2)$$

$$\bar{t} = t + \epsilon\beta(u,a,x,t) + O(\epsilon^2)$$

$$\bar{u} = u + \epsilon m_1(u,a,x,t) + O(\epsilon^2)$$

$$\bar{a} = a + \epsilon m_2(u,a,x,t) + O(\epsilon^2)$$

$$\bar{p_1} = p_1 + \epsilon P_1(u,a,x,t,p_1,p_2,q_1,q_2) + O(\epsilon^2)$$

$$\bar{p_2} = p_2 + \epsilon P_2(u,a,x,t,p_1,p_2,q_1,q_2) + O(\epsilon^2)$$

$$\bar{q_1} = q_1 + \epsilon Q_1(u,a,x,t,p_1,p_2,q_1,q_2) + O(\epsilon^2)$$

$$\bar{q_2} = q_2 + \epsilon Q_2(u,a,x,t,p_1,p_2,q_1,q_2) + O(\epsilon^2)$$

(4.78)

The functions α,β,m_1,m_2, P_1,P_2,Q_1 and Q_2 can be expressed in terms of two characteristic functions W_1 and W_2 as follows:

$$\alpha = \frac{\partial W_1}{\partial p_1} = \frac{\partial W_2}{\partial p_2}$$

$$\beta = \frac{\partial W_2}{\partial q_2} = \frac{\partial W_1}{\partial q_1}$$

$$m_1 = \frac{\partial W_1}{\partial p_1} p_1 + \frac{\partial W_1}{\partial q_1} q_1 - W_1$$

$$m_2 = \frac{\partial W_2}{\partial p_2} p_2 + \frac{\partial W_2}{\partial q_2} q_2 - W_2$$

$$-P_1 = \frac{\partial W_1}{\partial u} p_1 + \frac{\partial W_1}{\partial a} p_2 + \frac{\partial W_1}{\partial x}$$

$$-Q_1 = \frac{\partial W_1}{\partial u} q_1 + \frac{\partial W_1}{\partial a} q_2 + \frac{\partial W_1}{\partial t}$$

$$-P_2 = \frac{\partial W_2}{\partial u} p_1 + \frac{\partial W_2}{\partial a} p_2 + \frac{\partial W_2}{\partial x}$$

$$-Q_2 = \frac{\partial W_2}{\partial u} q_1 + \frac{\partial W_2}{\partial a} q_2 + \frac{\partial W_2}{\partial t}$$

(4.79)

The characteristics functions can be written as:

$$W_1(u, a, x, t, p_1, q_1) = W_{11}(u, a, x, t) p_1$$

$$+ W_{12}(u, a, x, t) q_1 + W_{13}(u, a, x, t) \tag{4.80a}$$

and

$$W_2(u, a, x, t, p_2, q_2) = W_{21}(u, a, x, t) p_2$$

$$+ W_{22}(u, a, x, t) q_2 + W_{23}(u, a, x, t) \tag{4.80b}$$

It can be seen that the characteristic functions are linear in p_1,p_2,q_1 and q_2.

Substituting the characteristic function from Eqs.(4.80) into the first four equations of Eq.(4.79), we get:

$$\alpha = W_{11} = W_{21} \;\; ; \;\; \beta = W_{12} = W_{22} \;\; ; \;\; m_1 = W_{13} \;\; ; \;\; m_2 = W_{23} \tag{4.81}$$

The characteristic functions can therefore be written in terms of α,β,m_1 and m_2 as:

$$W_1(u, a, x, t, p_1, q_1) = \alpha(u, a, x, t) p_1 + \beta(u, a, x, t) q_1$$

$$+ m_1(u, a, x, t)$$

and

$$W_2(u, a, x, t, p_2, q_2) = \alpha(u, a, x, t) p_2 + \beta(u, a, x, t) q_2$$

$$+ m_2(u, a, x, t) \tag{4.82b}$$

The characteristic functions in their new form are now substituted into the last four expressions of Eq.(4.79), leading to:

$$-P_1 = (\frac{\partial\alpha}{\partial u}p_1 + \frac{\partial\beta}{\partial u}q_1 - \frac{\partial m_1}{\partial u})p_1$$

$$+(\frac{\partial\alpha}{\partial a}p_1 + \frac{\partial\beta}{\partial a}q_1 - \frac{\partial m_1}{\partial a})p_2$$

$$+(\frac{\partial\alpha}{\partial x}p_1 + \frac{\partial\beta}{\partial x}q_1 - \frac{\partial m_1}{\partial x})$$

$$-P_2 = (\frac{\partial\alpha}{\partial u}p_2 + \frac{\partial\beta}{\partial u}q_2 - \frac{\partial m_2}{\partial u})p_1$$

$$+(\frac{\partial\alpha}{\partial a}p_2 + \frac{\partial\beta}{\partial a}q_2 - \frac{\partial m_2}{\partial a})p_2$$

$$+(\frac{\partial\alpha}{\partial x}p_2 + \frac{\partial\beta}{\partial x}q_2 - \frac{\partial m_2}{\partial x})$$

$$-Q_1 = (\frac{\partial\alpha}{\partial u}p_1 + \frac{\partial\beta}{\partial u}q_1 - \frac{\partial m_1}{\partial u})q_1$$

$$+(\frac{\partial\alpha}{\partial a}p_1 + \frac{\partial\beta}{\partial a}q_1 - \frac{\partial m_1}{\partial a})q_2$$

$$+(\frac{\partial\alpha}{\partial t}p_1 + \frac{\partial\beta}{\partial t}q_1 - \frac{\partial m_1}{\partial t})$$

$$-Q_2 = (\frac{\partial\alpha}{\partial u}p_2 + \frac{\partial\beta}{\partial u}q_2 - \frac{\partial m_2}{\partial u})q_1$$

$$+(\frac{\partial\alpha}{\partial a}p_2 + \frac{\partial\beta}{\partial a}q_2 - \frac{\partial m_2}{\partial a})q_2$$

$$+(\frac{\partial\alpha}{\partial t}p_2 + \frac{\partial\beta}{\partial t}q_2 - \frac{\partial m_2}{\partial t}) \tag{4.83}$$

Under the infinitesimal group of transformations, the system of differential equations, Eqs.(4.77), are invariant if

$$UG_i = 0 \quad (i = 1, 2) \tag{4.84}$$

or, in expanded form, we have

$$\alpha\frac{\partial G_i}{\partial x} + \beta\frac{\partial G_i}{\partial t} + m_1\frac{\partial G_i}{\partial u} + m_2\frac{\partial G_i}{\partial a}$$

$$+P_\mu\frac{\partial G_i}{\partial p_\mu} + Q_\nu\frac{\partial G_i}{\partial q_\nu} \tag{4.85}$$

where i,μ=1,2.

Substituting G_i from Eqs.(4.77) into Eq.(4.85), making use of the expressions given by Eq.(4.83) and eliminating q_1 and q_2, we get:

$$g_1 p_1 + g_2 p_2 + g_3 p_1^2 + g_4 p_2^2 + g_5 = 0 \qquad (4.86a)$$

and

$$g_6 p_1 + g_7 p_2 + g_8 p_1^2 + g_9 p_2^2 + g_{10} = 0 \qquad (4.86b)$$

where $g_1, \ldots, g_{10}$ are functions of p_1, p_2, q_1 and q_2 and are listed as follows:

$$g_1 = m_1 - u\frac{\partial \alpha}{\partial x} + \frac{2a}{k-1}\frac{\partial m_2}{\partial u} - \frac{\partial \alpha}{\partial t} + u^2\frac{\partial \beta}{\partial x}$$

$$+u\frac{\partial \beta}{\partial t} + a^2\frac{\partial \beta}{\partial x} - \frac{k-1}{2}a\frac{\partial m_1}{\partial a}$$

$$g_2 = \frac{2}{k-1}\left(m_2 + a\frac{\partial m_2}{\partial a} - a\frac{\partial \alpha}{\partial x} + au\frac{\partial \beta}{\partial x} - a\frac{\partial m_1}{\partial u}\right)$$

$$+a\frac{\partial \beta}{\partial t} + au\frac{\partial \beta}{\partial x}$$

$$g_3 = a^2\frac{\partial \beta}{\partial u} + \frac{k-1}{2}a\frac{\partial \alpha}{\partial a} - \frac{k-1}{2}au\frac{\partial \beta}{\partial a}$$

$$g_4 = -\frac{2a}{k-1}\frac{\partial \alpha}{\partial a} + \frac{2au}{k-1}\frac{\partial \beta}{\partial a} - \frac{4a^2}{(k-1)^2}\frac{\partial \beta}{\partial u}$$

$$g_5 = u\frac{\partial m_1}{\partial x} + \frac{2a}{k-1}\frac{\partial m_2}{\partial x} + \frac{\partial m_1}{\partial t}$$

$$g_6 = \frac{k-1}{2}\left(m_2 + a\frac{\partial m_1}{\partial u} - a\frac{\partial \alpha}{\partial x} + 2au\frac{\partial \beta}{\partial x}\right.$$

$$\left.-a\frac{\partial m_2}{\partial a} + \frac{\partial \beta}{\partial t}\right)$$

$$g_7 = m_1 + \frac{k-1}{2}a\frac{\partial m_1}{\partial a} - u\frac{\partial \alpha}{\partial x} - \frac{\partial \alpha}{\partial t} + a^2\frac{\partial \beta}{\partial x}$$

$$-\frac{2a}{k-1}\frac{\partial m_2}{\partial u} + u^2\frac{\partial \beta}{\partial x} + u\frac{\partial \beta}{\partial t}$$

$$g_8 = \frac{k-1}{2}\left(-a\frac{\partial \alpha}{\partial u} + au\frac{\partial \beta}{\partial u} - \frac{k-1}{2}a^2\frac{\partial \beta}{\partial a}\right)$$

$$g_9 = a^2\frac{\partial \beta}{\partial a} + \frac{2a}{k-1}\frac{\partial \alpha}{\partial u} - \frac{2a}{k-1}u\frac{\partial \beta}{\partial u}$$

$$g_{10} = \frac{k-1}{2}a\frac{\partial m_1}{\partial x} + u\frac{\partial m_2}{\partial x} + \frac{\partial m_2}{\partial t}$$

Since these functions are independent of the p's and q's, all the g's must be zero. Putting these g's to zero, rearranging the terms and combining the equations, we finally get:

$$u\frac{\partial\beta}{\partial u} - \frac{\partial\alpha}{\partial u} - \frac{k-1}{2}a\frac{\partial\beta}{\partial a} = 0$$

$$a\frac{\partial\beta}{\partial u} - \frac{k-1}{2}u\frac{\partial\beta}{\partial a} + \frac{k-1}{2}\frac{\partial\alpha}{\partial a} = 0$$

$$\frac{\partial m_1}{\partial u} = \frac{\partial m_2}{\partial a}$$

$$\frac{2}{k-1}\frac{\partial m_2}{\partial u} - \frac{k-1}{2}\frac{\partial m_1}{\partial a} = 0$$

$$u\frac{\partial m_1}{\partial x} + \frac{\partial m_1}{\partial t} + \frac{2a}{k-1}\frac{\partial m_2}{\partial x} = 0$$

$$a\frac{\partial m_1}{\partial x} + \frac{2u}{k-1}\frac{\partial m_2}{\partial x} + \frac{2}{k-1}\frac{\partial m_2}{\partial t} = 0$$

$$m_1 - \frac{\partial\alpha}{\partial t} + u\frac{\partial\beta}{\partial t} - u\frac{\partial\alpha}{\partial x} + (u^2 + a^2)\frac{\partial\beta}{\partial x} = 0$$

$$m_2 + a(\frac{\partial\beta}{\partial t} - \frac{\partial\alpha}{\partial x}) + 2au\frac{\partial\beta}{\partial x} = 0$$

(4.87)

Any form of the functions α,β,m_1 and m_2 satisfying all eight equations in Eq.(4.87) will be a group of transformations.

Consider the special case in which both m_1 and m_2 are linear with respect to u and a. Thus, we write:

$$m_1 = m_{11}(t,x)u + m_{12}(t,x)a + m_{13}(t,x)$$

$$m_2 = m_{21}(t,x)u + m_{22}(t,x)a + m_{23}(t,x)$$

(4.88)

Substituting the above into the third and fourth equation of Eq.(4.87) we get:

$$m_{22} = m_{11} \quad ;\cdot \quad m_{21} = \frac{(k-1)^2}{4}m_{12}$$

The functions m_1 and m_2 therefore becomes:

$$m_1 = m_{11}(t,x)u + m_{12}(t,x)a + m_{13}(t,x) \tag{4.89a}$$

$$m_2 = \frac{(k-1)^2}{4}m_{12}(t,x)u + m_{11}(t,x)a + m_{23}(t,x) \tag{4.89b}$$

from which, the fifth and sixth equations of Eq.(4.87), we get:

$$\frac{\partial m_{13}}{\partial t}+\frac{\partial m_{11}}{\partial x}u^2+\frac{\partial m_{12}}{\partial x}\Big(\frac{k+1}{2}\Big)au+\Big(\frac{\partial m_{13}}{\partial x}+\frac{\partial m_{11}}{\partial t}\Big)u$$

$$+\Big(\frac{\partial m_{12}}{\partial t}+\frac{2}{k-1}\frac{\partial m_{23}}{\partial x}\Big)a+\frac{2}{k-1}\Big(\frac{\partial m_{11}}{\partial x}\Big)a^2=0 \tag{4.90a}$$

and

$$\frac{\partial m_{23}}{\partial t}+\Big(\frac{\partial m_{13}}{\partial x}+\frac{\partial m_{11}}{\partial t}\Big)a+\Big(\frac{k-1}{2}\frac{\partial m_{12}}{\partial t}+\frac{\partial m_{23}}{\partial x}\Big)u$$

$$+\Big(\frac{k+1}{k-1}\frac{\partial m_{11}}{\partial x}\Big)au+\frac{\partial m_{12}}{\partial x}a^2+\frac{k-1}{2}\frac{\partial m_{12}}{\partial x}u^2=0 \tag{4.90b}$$

Eqs.(4.90) are satisfied identically only if all the coefficients are zero. We then get:

$$\frac{\partial m_{13}}{\partial t}=\frac{\partial m_{23}}{\partial t}=\frac{\partial m_{11}}{\partial x}=\frac{\partial m_{12}}{\partial x}=0 \tag{4.91a}$$

$$\frac{\partial m_{13}}{\partial x}+\frac{\partial m_{11}}{\partial t}=0 \tag{4.91b}$$

$$\frac{\partial m_{12}}{\partial t}+\Big(\frac{2}{k-1}\Big)\frac{\partial m_{23}}{\partial x}=0 \tag{4.91c}$$

Eq.(4.91a) shows that m_{13} and m_{23} are functions of x only, and m_{11} and m_{12} are functions of t only. Let us put

$$m_{13}=C_{11}x+C_{12} \tag{4.92a}$$

$$m_{23}=A_0x+B_0 \tag{4.92b}$$

Substituting Eq.(4.92) into Eqs.(4.91b,c), we get

$$m_{11}=-C_{11}t+C_{14} \tag{4.93a}$$

$$m_{12}=-\frac{2}{k-1}A_0t+C_2 \tag{4.93b}$$

Substituting m_{11},m_{12},m_{13} and m_{23} into Eq.(4.88), we then get:

$$m_1=(-C_{11}t+C_{14})u-\Big(\frac{2}{k-1}A_0t-C_2\Big)a+(C_{11}x+C_{12})$$

$$m_2=\frac{(k-1)^2}{4}\Big(-\frac{2}{k-1}A_0t+C_2\Big)u+\Big(-\frac{k-1}{2}C_{11}t+C_{14}\Big)a$$

$$+(A_0x+B_0) \tag{4.94a, b}$$

For the functions α and β, let us consider the case in which

$$\alpha(t,x,u,a) = \alpha_2 x + \alpha_1 t + \alpha_0(u,a) \tag{4.95a}$$

$$\beta(t,x,u,a) = \beta_2 x + \beta_1 t + \beta_0(u,a) \tag{4.95b}$$

Substitution of α and β from Eq.(4.95) into the first two equations of Eq.(4.87) leads to

$$u\frac{\partial \beta_0}{\partial u} - \frac{\partial \alpha_0}{\partial u} - \frac{k-1}{2} a \frac{\partial \beta_0}{\partial a} = 0 \tag{4.96a}$$

$$a\frac{\partial \beta_0}{\partial u} - \frac{k-1}{2} u \frac{\partial \beta_0}{\partial a} + \frac{k-1}{2}\frac{\partial \alpha_0}{\partial a} = 0 \tag{4.96b}$$

Next, α and β are substituted into the last two equations of Eq.(4.87) and we get:

$$(C_{12} - \alpha_1) + C_{11}x + (C_{14} + \beta_1 - \alpha_2)u - C_{11}t - \frac{2}{k-1}A_0 at$$

$$-C_2 a + C_{11}x + \beta_2(u^2 + a^2) = 0 \tag{4.97a}$$

$$B_0 + A_0 x - \left(\frac{k-1}{2}A_0\right)ut + \frac{(k-1)^2}{4}C_2 u - C_{11}ta$$

$$+(C_{14} + \beta_1 - \alpha_2)a + 2\beta_2 au = 0 \tag{4.97b}$$

For Eqs.(4.97) to be satisfied identically, we get:

$$C_{11} = A_0 = C_2 = \beta_2 = B_0 = 0$$

$$C_{12} - \alpha_1 = 0 \tag{4.98}$$

$$C_{14} + \beta_1 - \alpha_2 = 0$$

Using Eq.(4.98), the final form of α, β, m_1 and m_2 can be written as:

$$\alpha = \alpha_2 x + C_{12}t + \alpha_0(u,a)$$

$$\beta = (\alpha_2 - C_{14})t + \beta_0(u,a)$$

$$m_1 = C_{14}u + C_{12}$$

$$m_2 = C_{14}a$$

which are identical to the group of transformations derived by Mueller and Matschat[10].

4.9 Summary

In this chapter, both direct as well as group-theoretic methods have been applied to a variety of problems in engineering science. The direct methods such as the separation of variables and dimensional analysis are relatively simple to apply. The analysis, however, is restrictive because of the assumed form of transformations. The group-theoretic methods such as the Birkhoff-Morgan method and the Hellums- Churchill method utilize the underlying idea of invariance of the equations under an assumed group of transformations. Again, the resulting invariant solutions would be restrictive. Most boundary value problems of physical interest are invariant under linear and spiral groups of transformations. The group invariants for these groups can be readily obtained by solving a set of algebraic equations that arise as a result of invoking invariance of the partial differential equations and auxiliary conditions. From an engineering standpoint, the advantage of the Hellums-Churchill method is that the notion of "dimensions" are implied in the procedure and the final representation obtained is either self-similar or normalized. The deductive group procedures based on a general infinitesimal group of transformations lead to invariant solutions not obtainable by either Birkhoff-Morgan, Hellums-Churchill or any other direct or inspectional group procedures.

REFERENCES

1. Abbott,D.E. and Kline,S.J.,Simple Methods for Construction of Similarity Solutions of Partial Differential Equations, AFOSR TN 60-1163, Report MD-6, Dept. of Mech. Eng., Stanford Univ. (1960).
2. Pai,S.I., Viscous Flow Theory, Vol.1, p.62, D. Van Nostrand Company, Inc.,Princeton,N.J.(1960).
3. Taulbee,D.R.,Cozzarelli,F.A. and Dym,C.L.,"Similarity Solutioins to Some Nonlinear Impact Problems", Int'l J. of Nonlinear Mechanics, Vol.6 (1971).
4. Chand,R., Davy,D.T. and Ames,W.F.,"On the Similarity Solutions for a General Class of Nonlinear Dissipative Materials",Int'l J. of Nonlinear Mechanics, Vol.11 (1976).
5. Van Dyke, M.,"Free Convection From a Vertical Needle", Problems of Hydrodynamics and Continuum Mechanics", S.I.A.M. (1969).
6. Yih,C.S.,"Free Convection Due to a Point Source of Heat",Proc. lst U.S. National Congress of Applied Mechanics (1951).
7. Morgan, A.J.A.,"The Reduction of One of the Number of Independent Variables in Some Systems of Partial Differential Equations",Quar. J. of Math., Oxford (2) (1952).
8. Kalthia,N.L. and Jain,R.K.,"Similarity Solutions of the Heated Jet With Variable Viscosity and Thermal Conductivity ", Symmetry, Similarity and Group Theoretic Methods in Mechanics, Calgary, Canada (1974).

9. Na,T.Y.,"Group-Theoretic Analysis of Unsteady One-Dimensional Gas Dyn. Eqs.",Symmetry, Similarity and Group Theoretic Methods in Mechanics, Calgary, Canada (1974).
10. Mueller, von Ernst-August and Matschat, Klaus, "Uber das Auffinden von Ahnlichkeitslosungen Partieller Differential Gleichungs Systeme unter Benutzung von Transformations Gruppen, mit Anwendungen auf Probleme der Stromungsphysik, Akademie-Verlag, Berlin (1962).

Chapter 5

SIMILARITY ANALYSIS OF BOUNDARY VALUE PROBLEMS WITH FINITE BOUNDARIES

5.0 Introduction

It is commonly believed that similarity analysis of boundary value problems in science and engineering is domain-and-boundary condition limited, in that semi-infinite or infinite domains are required. A review of literature on similarity would, indeed, reveal that similarity solutions are mostly available for boundary value problems that lack a characteristic length in one or more coordinate directions. In his book, Hansen [1] points out that problems with finite boundaries associated with finite, non-zero values do not usually possess similarity solutions. Therefore, he suggests that a lack of characteristic length in a coordinate direction could be used as a hint to proceed with similarity analysis and seek possible solutions.

A great majority of similarity solutions in science and engineering have been obtained by analysis based on dimensional or affine groups of transformations. This is understandable since most partial differential equations that represent physical problems are dimensionally homogeneous, and invariance of the equations under a dimensional group can be readily invoked.

A dimensional group of transformations can be expressed as

$$\bar{x}_i = a_1^{\gamma_{i1}} a_2^{\gamma_{i2}} \ldots\ldots . a_r^{\gamma_{ir}} x_i \tag{5.1}$$

$$(i = 1, 2, 3, \ldots .., m)$$

The one-parameter linear group of transformations is a special case of Eq.(5.1), and can be written as

$$\bar{x}_i = a^{\gamma_{i1}} x_i \tag{5.2}$$

$$(i = 1, 2, 3, \ldots ., m)$$

For further clarification, we consider the motion of a fluid over a suddenly accelerated infinite plate. The infinite plate is assumed to be immersed in an incompressible fluid which is at rest. At time $t = 0$, the plate is suddenly set in motion in its own plane at a constant velocity, U_0. The equation of motion is

$$\nu \frac{\partial^2 u}{\partial y^2} = \frac{\partial u}{\partial t} \tag{5.3}$$

where u is the fluid velocity parallel to the plate motion, and ν is the kinematic viscosity. The initial and boundary conditions are:

$$u(y, 0) = 0 \quad ; \quad u(0, t) = U_0 \quad ; \quad u(\infty, t) = 0 \tag{5.4a, b, c}$$

By invoking invariance of Eqs.(5.3) and (5.4) under a dimensional group of transformations, the similarity transformation can be written as

$$u(y,t) = U_0\phi(\varsigma) \quad where \quad \varsigma = \frac{y}{\sqrt{t}} \tag{5.5}$$

In the transformed coordinate, Eqs.(5.4) can be written as

$$\phi(\varsigma = 0) = 1 \;\; ; \;\; \phi(\varsigma \rightarrow \infty) = 0 \tag{5.6a, b}$$

When $y \rightarrow \infty$, the similarity variable $\varsigma \rightarrow \infty$. So is the case when y is finite and $t = 0$. Therefore, the boundary conditions

$$u(y,0) = u(\infty, t) = 0$$

would be the necessary consolidation of the auxiliary conditions that would lead to the requirement of semi-infinite domain in the direction of y. The need for semi-infinite or infinite domain is, for many physical problems, a direct consequence of invoking invariance under a dimensional or affine group of transformations. The questions that need to be raised are:(1) would it be possible to obtain similarity solutions for finite domain boundary value problems by seeking invariance under either dimensional or affine groups? (2) is it possible to obtain a broad classification of the type of boundary value problems for which similarity solutions would exist for finite domains? In the remainder of this chapter we examine, by way of examples, some instances for which similarity solutions in finite domains can be obtained.

More specifically, we consider the following types of problems:
(i) boundary value problems with moving boundaries,
(ii) invariant boundary and surface descriptions,
(iii)invariance under groups other than dimensional and affine.

5.1 Boundary Value Problems With Moving Boundaries [2]

Certain moving boundary problem descriptions become fixed when expressed in terms of similarity coordinates. The requirement for such problems is that the moving boundary should be invariant under the same group of transformations as are the governing equations and the other auxiliary conditions.

Consider the problem of freezing of a body of water of thickness D which is initially held at a temperature, T_f(phase change temperature, assumed to be zero degrees). Initially, the surface temperature drops to T_s and is subsequently held there. The surface $x = D$ is effectively insulated and the liquid temperature has a constant value, $T_f = 0$(see Figure 5.1).

The equation governing heat transfer in the frozen zone is given by

$$\frac{\partial^2 T_1}{\partial x^2} = \frac{1}{\alpha_1}\frac{\partial T_1}{\partial t} \qquad [0 < x < X(t)] \tag{5.7}$$

The boundary conditions are

$$T_1(0,t) = T_s \quad ; \quad T_1(X(t),t) = 0 \tag{5.8a, b}$$

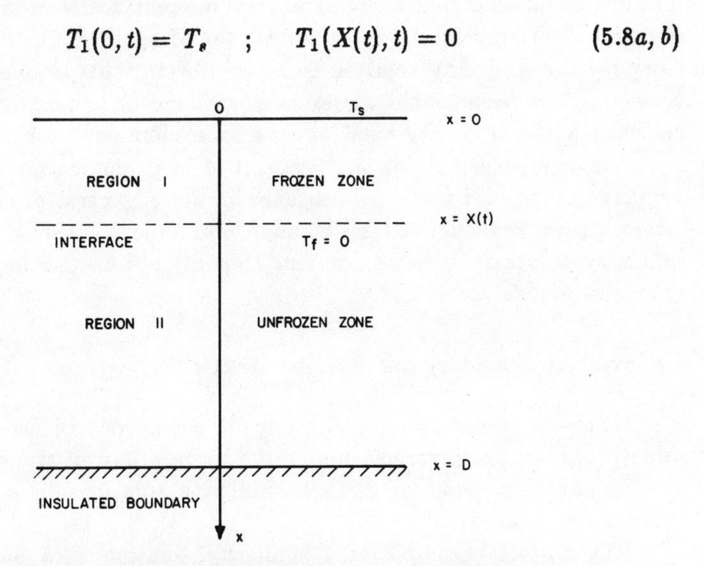

Fig.5.1 One-Dimensional Freezing Problem

At the moving boundary, the heat balance equation can be expressed as

$$K_1 \frac{\partial T_1}{\partial x} = \rho L \frac{dX}{dt} \tag{5.8c}$$

where α_1 is the thermal diffusivity, K_1 is the thermal conductivity, ρ is the density and L is the latent heat of fusion.

A similarity transformation can be sought using dimensional, or affine groups, and can be written as:

$$T_1(x,t) = T_s F(\varsigma), \quad where \ \varsigma = \frac{x}{2\sqrt{\alpha_1 t}} \tag{5.9}$$

The boundary conditions in the transformed coordinate are given by:

$$F(0) = 0 \quad ; \quad F(\beta) = 0 \tag{5.10a, b}$$

where $\varsigma = \beta$ locates the moving boundary such that

$$X(t) = 2\beta\sqrt{\alpha_1 t} \tag{5.11}$$

The value of β can be determined by introducing Eqs.(5.9) and (5.11) into Eq.(5.8c) to give the following relationship:

$$\beta e^{\beta^2} erf(\beta) + \frac{c_1 T_s}{\sqrt{\pi}} = 0 \tag{5.12}$$

c_1 is the specific heat of ice.

It can be seen that there is no requirement for semi-infinite domain, since Eq.(5.11) represents an invariant boundary. The finite domain of validity for the similarity solution is $0 < x < X(t)$. This is an example where invariance has been sought under a dimensional or affine group of transformations, and a similarity solution in a finite domain has been discovered.

Invariant solutions for the problem of heat conduction accompanying a change of phase based on invariance under a general group of transformations have been derived by Bluman and Cole[3]. These are examples of similarity solutions in finite domains that are obtained using groups other than dimensional or affine.

5.2 Invariant Boundary and Surface Description

Whenever the boundary surfaces or geometries of certain boundary value problems are invariant under the same group of transformations as are the governing equations, then similarity solutions in a finite domain could be discovered.

The simplest example of a similarity solution in a finite domain is the problem description with axisymmetry. Consider, the bending of an axisymmetric circular plates for which transverse deflection,w,is given by

$$\frac{\partial^4 w}{\partial x^4} + 2\frac{\partial^4 w}{\partial x^2 \partial y^2} + \frac{\partial^4 w}{\partial y^4} = \frac{q(x,y)}{D} \tag{5.13}$$

where $q(x, y)$ is the load per unit area, x and y are the Cartesian coordinates, and D is the bending rigidity of the plate. The symmetry is obvious and the coordinate transformation

$$r^2 = (x^2 + y^2) \tag{5.14}$$

would transform Eq.(5.13) into a description in terms of r alone as follows:

$$\frac{\partial^4 w}{\partial r^4} + \frac{2}{r}\frac{\partial^3 w}{\partial r^3} + \frac{1}{r^2}\frac{\partial^2 w}{\partial r^2} = \frac{q(r)}{D} \tag{5.15}$$

While transformations such as Eq.(5.14) convert a partial differential equation into an ordinary differential equation and are similarity transformations, they are seldom mentioned as such in similarity literature.

Consider now the problem of torsion of a conical shaft with a circular cross-section that is subject to terminal couples. The problem reduces to the solution of the following equation:

$$\frac{\partial^2 F}{\partial r^2} - \frac{3}{r}\frac{\partial F}{\partial r} + \frac{\partial^2 F}{\partial z^2} = 0 \tag{5.16}$$

where $F(r,z)$ is a stress function such that

$$\tau_{\theta z} = \frac{\mu}{r^2}\left(\frac{\partial F}{\partial r}\right) \tag{5.17}$$

and

$$\tau_{r\theta} = -\frac{\mu}{r^2}\left(\frac{\partial F}{\partial z}\right) \tag{5.18}$$

$\tau_{\theta z}$ and $\tau_{r\theta}$ are the shear stress components, and μ is the Lame's elastic constant. The cylindrical system of coordinates (r,θ,z) is used in the mathematical description (Figure 5.2).

The twisting moment, M, on any cross-section is given by

$$M = \int_0^a \tau_{\theta z} r(2\pi r dr)$$

$$= 2\pi\mu \int_0^a \frac{\partial F}{\partial r} dr \tag{5.19}$$

or,

$$M = 2\pi\mu[F(z,a) - F(z,0)] \tag{5.20}$$

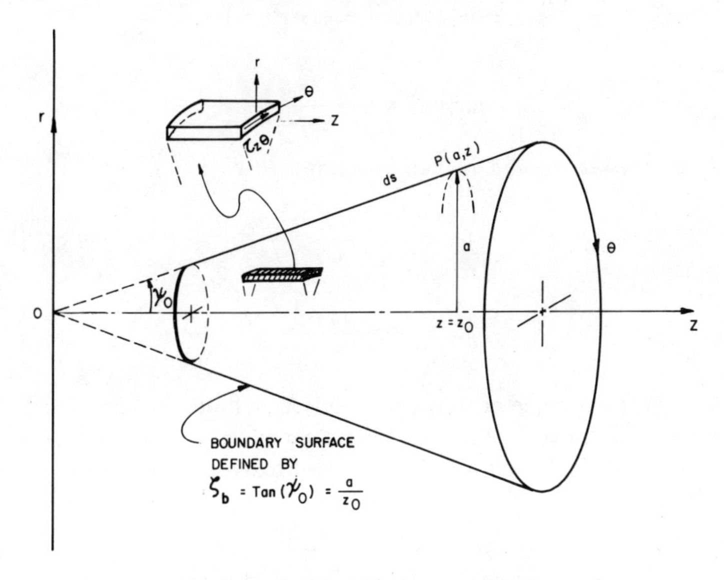

Fig.5.2 Torsion of Conical Shaft

The similarity transformation can be written as[4]

$$F(r,z) = C\phi(\varsigma), \quad where \ \varsigma = \frac{r}{z}. \tag{5.21}$$

The resulting ordinary differential equation can be written as

$$\varsigma(\varsigma^2+1)\frac{d^2\phi}{d\varsigma^2}+(2\varsigma^2-3)\frac{d\phi}{d\varsigma}=0 \tag{5.22}$$

Integrating Eq.(5.22) twice,

$$\phi(\varsigma)=A[\frac{1}{\sqrt{\varsigma^2+1}}-\frac{1}{3}(\frac{1}{\sqrt{\varsigma^2+1}})^3]+B \tag{5.23a}$$

or,

$$F(r,z)=\beta_0[\frac{z}{(r^2+z^2)^{1/2}}-\frac{1}{3}(\frac{z}{(r^2+z^2)^{1/2}})^3]+\beta_1 \tag{5.23b}$$

where A, B, β_0 and β_1. are integration constants.

The expression $z/(r^2+z^2)^{1/2}$ is a constant on the lateral surface of the conical section, since it is equal to the cosine of one half of the vertical angle of the conical shaft; hence, $F(r,z)$ would assume a constant value on the lateral surface.

The value of β_0 can be determined from Eq.(5.20) when the twisting couple in the terminal section is specified. Therefore,

$$\beta_0=\frac{M}{2\pi\mu\left[cos(\psi_0)-\frac{1}{3}cos^3(\psi_0)-\frac{2}{3}\right]} \tag{5.24}$$

where

$$cos(\psi_0)=\frac{z}{(r^2+z^2)^{1/2}}$$

The shear stresses $\tau_{\theta z}$ and $\tau_{r\theta}$ can be computed as

$$\tau_{\theta z}=-\frac{\mu\beta_0 rz}{(r^2+z^2)^{5/2}} \tag{5.25a}$$

$$\tau_{r\theta}=-\frac{\mu\beta_0 r^2}{(r^2+z^2)^{5/2}} \tag{5.25b}$$

As another example of similarity solution in finite domain descriptions, consider the classical problem of "spiral flows" of incompressible viscous fluids[5], as shown in Figure 5.3. The Navier-Stokes equations describing the flow can be written as:

$$(v_r\frac{\partial v_r}{\partial r}+\frac{v_\theta}{r}\frac{\partial v_r}{\partial\theta}-\frac{{v_\theta}^2}{r})$$

$$=-\frac{\partial p}{\partial r}+(\frac{\partial^2 v_r}{\partial r^2}+\frac{1}{r}\frac{\partial v_r}{\partial r}-\frac{v_r}{r^2}$$

$$+\frac{1}{r^2}\frac{\partial^2 v_r}{\partial\theta^2}-\frac{2}{r^2}\frac{\partial v_\theta}{\partial\theta}) \tag{5.26a}$$

$$\left(v_r\frac{\partial v_r}{\partial \theta}+\frac{v_\theta}{r}\frac{\partial v_\theta}{\partial \theta}+\frac{v_r v_\theta}{r}\right)$$

$$=-\frac{1}{r}\frac{\partial p}{\partial \theta}+\Big(\frac{\partial^2 v_\theta}{\partial r^2}+\frac{1}{r}\frac{\partial v_\theta}{\partial r}$$

$$-\frac{v_\theta}{r^2}+\frac{1}{r^2}\Big(\frac{\partial^2 v_\theta}{\partial \theta^2}\Big)+\frac{2}{r^2}\Big(\frac{\partial v_r}{\partial \theta}\Big)\Big) \tag{5.26b}$$

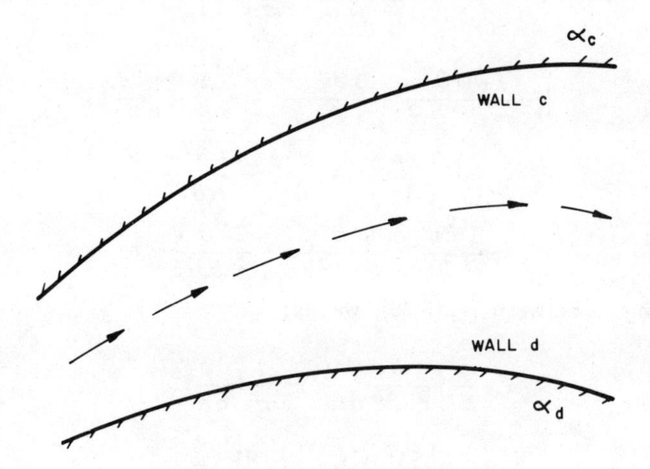

Fig.5.3 Spiral Viscous Flow

$$\frac{\partial v_r}{\partial r}+\frac{v_r}{r}+\frac{1}{r}\frac{\partial v_\theta}{\partial \theta}=0 \tag{5.26c}$$

The non-dimensional variables in Eq.(5.26) are related to their physical counterparts ($\bar{r}$,$\bar{v}_r$,$\bar{v}_\theta$,$\bar{p}$) by the following relationships:

$$r=\frac{\bar{r}}{L}\sqrt{Re}\ ,\quad v_r=\frac{\bar{v}_r}{U_0}\ ,\quad v_\theta=\frac{\bar{v}_\theta}{U_0}\sqrt{Re}$$

$$p=\frac{\bar{p}}{\rho U_0^2}\ ,\quad where\ \ Re=\frac{U_0 L}{\nu}$$

where $\bar{r}$ is the radial coordinate,$\bar{v}_r$ and $\bar{v}_\theta$ are the velocity components in the $\bar{r}$ and θ directions,$\bar{p}$ is the pressure, *Re* Reynolds number, ρ is the density, L is the is the length, ν is the kinematic viscosity and U_0 is a reference velocity.

We now introduce the stream function ψ as follows:

$$rv_r=\frac{\partial \psi}{\partial \theta}\ ,\quad v_\theta=-\frac{\partial \psi}{\partial r} \tag{5.27}$$

Introducing Eq.(5.27) into Eqs.(5.26) and simplifying, we obtain:

$$[\frac{1}{r^2}\frac{\partial\psi}{\partial\theta}\frac{\partial^2\psi}{\partial r\partial\theta} - \frac{1}{r^3}(\frac{\partial\psi}{\partial\theta})^2 + \frac{1}{r}\frac{\partial\psi}{\partial r}\frac{\partial^2\psi}{\partial\theta\partial r}$$

$$-\frac{1}{r}(\frac{\partial\psi}{\partial r})^2] = -\frac{\partial p}{\partial r} + [\frac{1}{r}\frac{\partial^3\psi}{\partial r^2\partial\theta}$$

$$+\frac{1}{r^2}\frac{\partial^2\psi}{\partial r\partial\theta} - \frac{1}{r^2}\frac{\partial^3\psi}{\partial\theta^2\partial r}] \tag{5.28a}$$

$$[-\frac{1}{r}\frac{\partial\psi}{\partial\theta}\frac{\partial^2\psi}{\partial r^2} + \frac{1}{r}\frac{\partial\psi}{\partial r}\frac{\partial^2\psi}{\partial\theta\partial r} - \frac{1}{r^2}\frac{\partial\psi}{\partial\theta}\frac{\partial\psi}{\partial r}]$$

$$= -\frac{1}{r}\frac{\partial p}{\partial\theta} + [-\frac{\partial^3\psi}{\partial r^3} - \frac{1}{r}\frac{\partial^2\psi}{\partial r^2}$$

$$+\frac{1}{r^2}\frac{\partial\psi}{\partial r} - \frac{1}{r^2}\frac{\partial^3\psi}{\partial\theta^2\partial r} + \frac{2}{r^2}\frac{\partial^3\psi}{\partial\theta^2\partial r}] \tag{5.28b}$$

Eliminating p between Eq.(5.28), we get:

$$\frac{\partial}{\partial\theta}[\frac{1}{r^2}\frac{\partial\psi}{\partial\theta}\frac{\partial^2\psi}{\partial r\partial\theta} - \frac{1}{r^3}(\frac{\partial\psi}{\partial\theta})^2$$

$$+\frac{1}{r}\frac{\partial\psi}{\partial\theta}\frac{\partial^2\psi}{\partial\theta\partial r} - \frac{1}{r}(\frac{\partial\psi}{\partial r})^2]$$

$$-\frac{\partial}{\partial r}[-\frac{\partial\psi}{\partial\theta}\frac{\partial^2\psi}{\partial r^2} + \frac{\partial\psi}{\partial r}\frac{\partial^2\psi}{\partial\theta\partial r} - \frac{1}{r}\frac{\partial\psi}{\partial\theta}\frac{\partial\psi}{\partial r}]$$

$$-\frac{\partial}{\partial\theta}[\frac{1}{r}\frac{\partial^3\psi}{\partial r^2\partial\theta} + \frac{1}{r^2}\frac{\partial^2\psi}{\partial r\partial\theta} - \frac{1}{r^2}\frac{\partial^3\psi}{\partial\theta^2\partial r}]$$

$$-\frac{\partial}{\partial r}[-r\frac{\partial^3\psi}{\partial r^3} - \frac{\partial^2\psi}{\partial r^2} + \frac{1}{r}\frac{\partial\psi}{\partial r} - \frac{1}{r}\frac{\partial^3\psi}{\partial\theta^2\partial r}$$

$$+\frac{2}{r}\frac{\partial^2\psi}{\partial\theta\partial r}] \tag{5.29}$$

Invocation of invariance of Eq.(5.29) under a spiral group of transformations

$$\theta^* = \theta + \alpha_1 a \ ; \ r^* = re^{\alpha_2 a} \ ; \ \psi^* = \psi e^{\alpha_3 a} \tag{5.30}$$

leads to $\alpha_1 = 0$,and the following similarity transformation:

$$\eta = \frac{r}{e^{\alpha\theta}} \ \ and \ \ \psi(\theta, r) = f(\eta) \tag{5.31}$$

The resulting ordinary differential equation becomes:

$$\alpha\eta[(\alpha^2 - \alpha)f'f'' - \frac{\alpha+1}{\eta}(f')^2]'$$

$$+\alpha(\alpha+\alpha^2)\eta[3\frac{f''}{\eta}+f'''+\frac{f'}{\eta^2}]'$$

$$+[(1+\alpha^2)\eta f'''+(1+2\alpha+3\alpha^2)f''$$

$$+(1+2\alpha+\alpha^2)\frac{f'}{\eta}]'=0 \tag{5.32a}$$

The boundary conditions are:

$$\eta=\eta_c\ :\ f(\eta_c)=c\ ,\ f'(\eta_c)=0 \tag{5.32b}$$

$$\eta=\eta_d\ :\ f(\eta_d)=d\ ,\ f'(\eta_d)=0 \tag{5.32c}$$

where

$$\eta_c=\frac{r}{e^{\alpha_c\theta}}\ ,\ \eta_d=\frac{r}{e^{\alpha_d\theta}}$$

The boundary descriptions

$$r=\eta_c e^{\alpha_c\theta}\quad and\quad r=\eta_d e^{\alpha_d\theta}$$

are invariant under Eq.(5.30) and would represent the walls through which the flow takes place. The similarity solution for the problem of spiral flow of an incompressible viscous fluid can therefore be obtained in finite domains by solving Eq.(5.32).

5.3 Invariance Under Groups Other Than Dimensional Groups

With the exception of physical problems involving a moving boundary,the restrictions of similarity solutions to semi-infinite or infinite domains can usually be attributed to the use of dimensional or affine groups of transformations. It is then reasonable to expect that such restrictions need not exist if invariance under groups other than dimensional or affine is sought.

Consider again, the linear heat equation

$$\frac{\partial^2 u}{\partial x^2}=\frac{\partial u}{\partial t} \tag{5.33}$$

We now define a spiral group of transformations

$$t=\bar{t}+\beta_1 a\ ;\quad x=e^{\beta_2 a}\bar{x}\ ;\quad u=e^{\beta_3 a}\bar{u} \tag{5.34}$$

For Eq.(5.33) to be invariant under the group defined above, β_2=0. The absolute invariants can be found by the Birkhoff-Morgan method as

$$\varsigma=x\ ;\quad u=e^{-n^2t}F_n(\varsigma),\qquad where\ n^2=-\frac{\beta_3}{\beta_1} \tag{5.35}$$

Eq.(5.35) corresponds to the classical separation of variables, $u = X(x)\,T(t)$, which can be used to obtain eigen-value solutions in finite domains.

Example 5.1: Consider a finite bar of length L with a heat source at $x = x_0$ $(0 < x_0 < L)$. Each end of the bar is held at a temperature of zero.

The function $F_n(\varsigma)$ in Eq.(5.35) satisfies the ordinary differential equation

$$\frac{d^2F_n}{dx^2} + n^2F_n = 0 \tag{5.36}$$

Therefore,

$$F_n(x) = A_n\ sin(nx) + B_n\ cos(nx) \tag{5.37}$$

Using the boundary condition $u(0,t) = 0$, we get $B_n = 0$. The eigenvalues are obtained by applying the condition $u(L,t) = 0$,giving

$$A_n\ sin(nL) = 0$$

Therefore,

$$nL = m\pi \quad (m = 1, 2, 3, ..., \infty)$$

The solution can now be written as

$$u(x,t) = \sum_{n=1}^{\infty} u_n(x,t)$$

$$= \sum_{m=1}^{\infty} A_m\ sin(\frac{m\pi x}{L})\,exp(-\frac{m^2\pi^2 t}{L^2} \tag{5.38}$$

If the initial condition is specified as

$$u(x,0) = \delta(x - x_0)\ ,\ then\ \ A_m = \frac{2}{L} sin(\frac{m\pi x_0}{L}).$$

The required temperature distribution for the finite bar is given by

$$u(x,t) = \sum_{m=1}^{\infty} \frac{2}{L} sin(\frac{m\pi x_0}{L})\,sin(\frac{m\pi x}{L})\,exp(-\frac{m^2\pi^2 t}{L^2} \tag{5.39}$$

The same arguments can be used for the linear wave equation governing transverse oscillations of a string, which can be written as:

$$c^2\frac{\partial^2 w}{\partial x^2} = \frac{\partial^2 w}{\partial t^2} \tag{5.40}$$

where w is the transverse displacement and c is a constant. Defining a group of transformations

$$t = \bar{t} + \gamma_1 a\ \ ;\ \ x = \bar{x}e^{\gamma_2 a}\ \ ;\ \ w = \bar{w}e^{\gamma_3 a} \tag{5.41}$$

we find that Eq.(5.40) is invariant if $\gamma_2 = 0$.

The absolute invariants are given by

$$\xi = x \quad and \quad w_n = e^{-\lambda^2 t}\phi(\xi), \quad where \ \lambda^2 = -\frac{\gamma_3}{\gamma_1} \tag{5.42}$$

If γ_3/γ_1 is set equal to $i\omega$, then the solution can be written as:

$$w = \sum_{i=1}^{\infty} e^{i\omega_n t}\phi_n(x) \tag{5.43}$$

$\phi_n(x)$ corresponds to the eigen functions and ω_n corresponds to the eigen frequencies for a given finite domain boundary value problem. Methods for finding these eigenvalues are discussed in text on vibrations[6]. Again, there is no requirement for semi-infinite or infinite domains.

5.4 Summary

Traditional similarity solutions in science and engineering have been mostly obtained by seeking invariance of the governing equations and auxiliary conditions under either a dimensional or an affine group of transformations. The requirement for semi- infinite or infinite domains has been a consequence of a need for a consolidation of auxiliary conditions in order that a similarity representation is possible for such group invariances.

By way of examples, it was shown that similarity solutions for finite domain descriptions are possible for the following types of problems:

(1) boundary value problems with moving boundaries for which similarity solutions can be obtained by seeking invariance under any group of transformations including dimensional and affine,
(2) domains with invariant boundary or surface descriptions, and
(3) boundary value problems for which invariance is sought under groups of transformations other than dimensional and affine.

It can be concluded that similarity solutions for finite domain boundary value problems are possible if the governing equations and their associated auxiliary conditions are invariant under an appropriate group of transformations. Semi-infinite and infinite domains are mainly a consequence of invoking invariance under a dimensional or an affine group of transformations.

REFERENCES

1. Hansen,A.G.,Similarity Analysis of Boundary Value Problems in Engineering, Princeton Hall (1960).

2. Carslaw,H.S. and Jaeger,J.C.,Conduction of Heat in Solids, 2nd edition,Oxford Univ. Press,London (1959).
3. Bluman, G.W.,"Applications of the General Similarity Solution of the Heat Equation to Boundary Value Problems", Quar. Appl. Math., Vol.31, No. 4 (1974).
4. Seshadri, R.,Group Theoretic Methods of Similarity Analysis Applied to Nonlinear Impact Problems, Ph.D. thesis, Univ. of Calgary, Alberta,Canada (1973).
5. Birkhoff, G., Hydrodynamics, Princeton University Press, New Jersey (1960).
6. Nowacki, W.,Dynamics of Elastic Systems, Chapman and Hall Ltd., London (1963).

Chapter 6

ON OBTAINING NON-SIMILAR SOLUTIONS FROM SIMILAR SOLUTIONS

6.0 Introduction

We have seen in the earlier chapters, that a similarity representation is obtainable for a boundary value problem provided the governing differential equations and the associated boundary conditions are invariant under a group of transformations. However, if any of the equations and boundary conditions is not invariant under a group, then the problem becomes nonsimilar.

Attempts have been made in the past to generate solutions for non-similar problems from known similarity solutions [1]. These techniques have mainly been applied to linear equations for which other well-known techniques such as integral transforms are available. In this chapter, different methods for obtaining non-similar solutions from similar solutions for linear as well as nonlinear equations are examined.

6.1 Superposition of Similarity Solutions

When a given linear partial differential equation system is "almost" invariant (some term in the equations and/or auxiliary conditions is not invariant under a continuous group of transformations), new solutions to the non-similar problem can be constructed by using a linear combination of similarity solutions[2].

Example 6.1: Stokes Second Problem

For the purpose of illustration of the method, we will consider the classical case of Stokes' second problem in which a plate oscillates harmonically parallel to itself at a velocity of $U_0 \cos(\omega t)$. The velocity of the fluid above the plate can be found by solving the simplified Navier-Stokes equation:

$$\nu \frac{\partial^2 u}{\partial y^2} = \frac{\partial u}{\partial t} \tag{6.1}$$

subject to the boundary conditions:

$$u(0,t) = U_0 cos(\omega t) \tag{6.2a}$$

$$u(\infty,t) = 0 \tag{6.2b}$$

and the initial condition

$$u(y,0) = 0 \tag{6.2c}$$

Eqs.(6.1),(6.2b) and (6.2c) are invariant under a one-parameter linear transformation group, G_1:

$$G_1 \ : \ \bar{u} = \alpha^{2m} u \ ; \ \bar{x} = \alpha x \ ; \ \bar{t} = \alpha^2 t \tag{6.3}$$

where the constant α is the parameter of transformation. The boundary condition,Eq.(6.2a), is not invariant under the above group. Therefore, a similarity solution cannot be directly determined for the problem. Since Eq.(6.1) is linear, the method of superposition can be applied.

The similarity variable can be written as

$$\varsigma = \frac{y}{2\sqrt{\nu t}} \ ; \ F_n(\varsigma) = \frac{u}{c_n t^n} \tag{6.4}$$

The similarity representation is given by:

$$F_n'' + 2\varsigma F_n' - 4nF_n = 0 \tag{6.5a}$$

and

$$F_n(\infty) = 0 \tag{6.5b}$$

Since Eq.(6.2a) is not invariant under group G_1, the principle of superposition can be utilized as follows:

$$u(y,t) = \sum_{n=0}^{\infty} c_n t^n F_n(\varsigma) \tag{6.6}$$

The boundary condition, Eq.(6.2a), can be written as

$$\frac{u(0,t)}{U_0} = cos(\omega t) = \sum_{n=0}^{\infty} \frac{(\omega t)^n}{n!}(-1)^{n/2} \tag{6.5c}$$

$$(n = 0, 2, 4, ...)$$

It follows that

$$c_n = U_0 \frac{\omega^n}{n!}(-1)^{n/2} \tag{6.6a}$$

$$(n = 0, 2, 4, ...)$$

and

$$F_n(0) = 1 \tag{6.6b}$$

Therefore, the solution can be expressed as

$$\frac{u(x,t)}{U_0} = \sum_{n=0}^{\infty} \frac{(\omega t)^n}{n!}(-1)^{n/2} F_n(\varsigma) \tag{6.7a}$$

$$(n = 0, 2, 4, ...)$$

$$= F_0(\varsigma) - \frac{(\omega t)^2}{2!}F_2(\varsigma) + \frac{(\omega t)^4}{4!}F_4(\varsigma) + ... \tag{6.7b}$$

where the functions $F_n(\varsigma)$ are obtained by solving Eq.(6.5a),subject to the boundary conditions

$$F_n(0) = 1 \quad ; \quad F_n(\infty) = 0 \tag{6.8}$$

Na and Hansen[2] have expressed the solution for Eq.(6.5a) subject to boundary conditions,Eq.(6.8), in terms of Hermite functions as follows:

$$F_n(\varsigma) = \frac{H_{2n}(i\varsigma)}{H_{2n}(0)}\left(1 - \frac{1}{I}\int_0^{\varsigma} \frac{exp(-\eta^2)\,d\eta}{H_{2n}^2(i\eta)}\right) \tag{6.9a}$$

where the Hermite function is defined as:

$$H_{2n}(i\varsigma) = \sum_{k=0}^{n} \frac{(-1)^n 2n(2\varsigma)^{2n-2k}}{k!(2n-2k)!} \tag{6.9b}$$

and

$$I = \int_0^{\infty} \frac{exp(-\eta^2)\,d\eta}{H_{2n}^2(i\eta)} \tag{6.9c}$$

In order to show the accuracy of the superposed similarity solutions, we will now compare it with the exact solution:

$$\frac{U(x,t)}{U_0} = e^{-\sqrt{2\omega t}\varsigma} cos(\omega t - \sqrt{2\omega t}\varsigma) \tag{6.10}$$

It can be seen from figure 6.1 that good agreement is obtained.

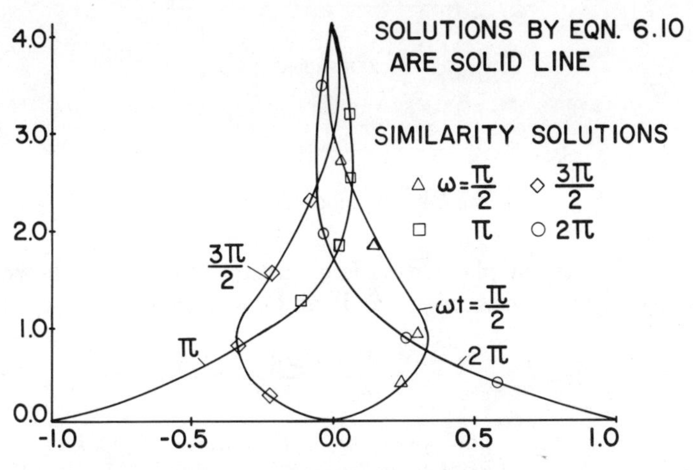

Fig. 6.1 Comparison of Two Solutions

Example 6.2 One-dimensional Consolidation of Thawing Soils

As another illustration, consider the problem of consolidation of thawing soils. Permafrost soils provide adequate bearing capacity for supporting most structures as long as soils remain in a frozen state. The thawing of permafrost soils which contain large quantities of ice can cause severe engineering problems. When frozen soils thaw, water is released and settlement would occur as the water is expelled from the ground. If the rate of water generation exceeds the discharge capacity of the soil, excess pore pressure will develop resulting in a reduction of shear strength of the soil, and subsequent failures of slopes and foundations.

A one-dimensional configuration (Figure 6.2) is considered for a step increase in surface temperature in the semi-infinite homogeneous mass of frozen ground. The movement of the thaw front is given by:

$$X(t) = \alpha\sqrt{t} \tag{6.11}$$

where α is a constant, t is time and $X(t)$ is the distance of the thaw boundary from the surface. It is assumed that the soil is compressible in the thawed zone.

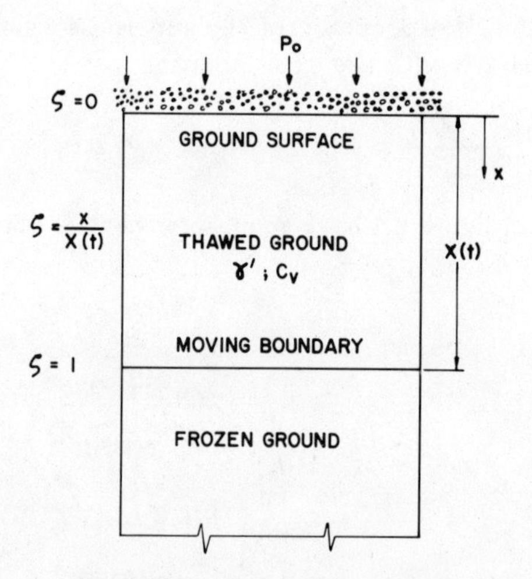

Fig. 6.2 One-Dimensional Thaw Consolidation

The consolidation phenomena is described by the well- known Terzaghi's equation of soil mechanics[3] and can be written as:

$$c_v \frac{\partial^2 u}{\partial x^2} = \frac{\partial u}{\partial t} \tag{6.12}$$

where u is the excess pore pressure in the thawed soil, x is the depth below surface, and c_v is the coefficient of consolidation. Eq.(6.12) is valid for $0 < x < X(t)$ and for $t > 0$. Since the surface $x = 0$ is considered free-draining, the excess pore pressure

$$u(x=0,t) = 0 \quad t > 0 \tag{6.13a}$$

The water that is expelled at the moving boundary during thawing flows outwards if an excess pore pressure is developed at $x = X(t)$. For a saturated soil in which the effective stress-strain law is linear and Darcy's flow law is valid, the moving boundary condition is given by:

$$p_0 + \gamma' X(t) - u(X,t) = c_v \frac{\partial u(X(t))}{\partial x}\left(\frac{dX}{dt}\right)^{-1} \tag{6.13b}$$

where γ'=density of soil minus density of water, and p_0 is the applied load on the surface.

The initial condition is given by

$$u(x, t=0) = 0 \tag{6.13c}$$

For $X(t) = \alpha\sqrt{t}$, an examination of Eqs.(6.12) and (6.13b) would reveal that a direct similarity transformation is not possible. A superposition of similarity transformations, however, would result in a meaningful solution. Therefore,

$$u(x,t) = c_1 f_1(\varsigma) + c_2\sqrt{t} f_2(\varsigma) \tag{6.14}$$

where the similarity variable is

$$\varsigma = \frac{x}{\alpha\sqrt{t}}$$

and the constants are

$$c_1 = p_0 \quad and \quad c_2 = \gamma'\alpha$$

The similarity representation can be written as follows:

$$\frac{\partial^2 f_1}{\partial \varsigma^2} + 2R^2\varsigma\frac{df_1}{d\varsigma} = 0 \tag{6.15}$$

with

$$f_1(0) = 0 \quad and \quad 2R^2[f_1(\varsigma=1) - 1] = \frac{d\, f(\varsigma=1)}{d\varsigma} \tag{6.16a, b}$$

where $R = \alpha/(2\sqrt{c_v})$ is the thaw-consolidation ratio.

Eqs.(6.15) and (6.16) can be solved to give

$$f_1(\varsigma) = \frac{erf(R\varsigma)}{erf(R) + exp(-R^2)/(\sqrt{\pi}R)} \tag{6.17}$$

where erf( ) is an error function.

Similarly, $f_2(\varsigma)$ can be obtained by solving the following:

$$\frac{d^2 f_2}{d\varsigma^2} + 2R^2\varsigma\frac{df_2}{d\varsigma} - 2R^2 f_2 = 0 \tag{6.18}$$

$$f_2(0) = 0 \;\; and \;\; 2R^2[f_2(\varsigma = 1) - 1] = \frac{df_2(\varsigma = 1)}{d\varsigma} \tag{6.19}$$

The solution for Eq.(6.18) can be written as:

$$f_2(\varsigma) = exp[-(R^2\varsigma^2/2)][c_1 U(\frac{3}{2}, R\varsigma\sqrt{2}) + c_2 V(\frac{3}{2}, R\varsigma\sqrt{2})] \tag{6.20}$$

where $U(\;)$ and $V(\;)$ are parabolic cylinder functions. On applying the boundary conditions,Eq.(6.19), and after carrying out some manipulations, we get:

$$f_2(\varsigma) = \frac{\varsigma}{1 + (.5R^{-2})} \tag{6.21}$$

The superposed solution can therefore be written as

$$u(x,t) = u_1(x,t) + u_2(x,t) \tag{6.22}$$

where

$$u_1(x,t) = \frac{p_0 erf(R\varsigma)}{R\sqrt{\pi}erf(R) + exp(-R^2)}\sqrt{\pi}R$$

and

$$u_2(x,t) = \frac{\gamma' x}{1 + 1/(2R^2)}$$

Eq.(6.22) describes the variation of excess pore pressure in a thawing soil.

6.2 The Use of Fundamental Solutions

We will now consider a partial differential equation of the form:

$$Lu(x,t;x_0,t_0) = \delta(x - x_0)\delta(t - t_0) \tag{6.23}$$

where L is a self-adjoint differential operator, and (x_0, t_0) are the poles[4]. The notation $u\;(x,t;x_0,t_0)$ is introduced to emphasize the dependence of the solution on the poles x_0 and t_0. Any solution of Eq.(6.23) is termed as a "fundamental solution". While the general fundamental solution contains an arbitrary number of constants, the so-called Green's function is obtained once the boundary conditions are specified.

Example 6.3: Source Solution for Linear Heat Conduction

Consider a one-dimensional heat conduction process in an infinite bar with constant thermal properties. The bar is initially maintained at a uniform temperature (see Fig.6.3). The governing equation for temperature in the bar can be written as:

$$\rho c \frac{\partial T}{\partial t} - k \frac{\partial^2 T}{\partial x^2} = Q\,\delta(x - x_0)\,\delta(t - t_0) \tag{6.24}$$

In the above description, the source of heat, Q,is located at a distance x_0 from the origin. k is the thermal conductivity,ρ is the density of the bar, and c is the specific heat.

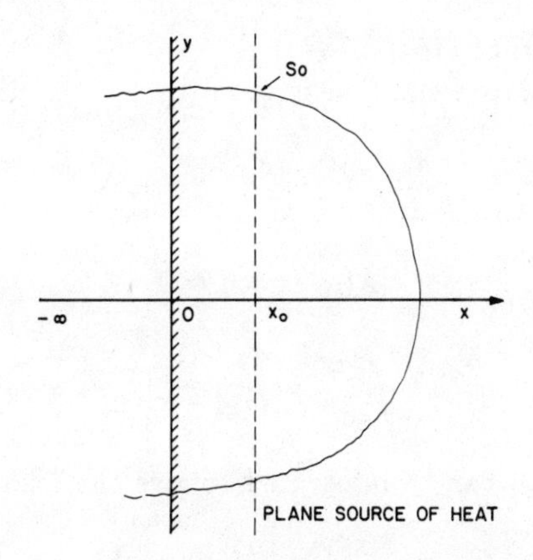

Fig. 6.3 Conduction With Heat Source

Making a change of variables $\bar{x} = x - x_0$ and $\bar{t} = t - t_0$, the Green's Function for the heat conduction problem is the solution of:

$$(\frac{\partial}{\partial \bar{t}} - \lambda \frac{\partial^2}{\partial \bar{x}^2})\,g(\bar{x}, \bar{t}) = \frac{Q}{\rho c}\delta(\bar{x})\delta(\bar{t}) \tag{6.25}$$

$$(-\infty < \bar{x} < \infty\ ;\ \bar{t} \geq 0)$$

λ is the thermal conductivity,equal to $k/(\rho c)$. The auxiliary conditions are:

$$T(\bar{x}; \bar{t} = 0) \quad ; \quad T(\bar{x} \rightarrow \pm\infty; \bar{t}) = 0 \tag{6.26}$$

In addition, the law of conservation of total heat, which can be written as:

$$\rho c \int_{-\infty}^{+\infty} T(\bar{x}, \bar{t})\,d\bar{x} = Q \tag{6.26a}$$

must be satisfied.

Eqs.(6.25) and (6.26) are invariant under a group of transformations

$$T^* = \beta^{-1/2}\bar{T} \;\; ; \;\; x^* = \beta^{1/2}\bar{x} \;\; ; \;\; t^* = \beta\bar{t} \tag{6.27}$$

Using any of the techniques described in chapter 3, the similarity transformation can be written as

$$g(\bar{x},\bar{t}) = \frac{Q}{\rho c\sqrt{\lambda\bar{t}}}F(\lambda) \tag{6.28}$$

where

$$\varsigma = \frac{\bar{x}}{2\sqrt{\lambda\bar{t}}}$$

It should be noted that $\delta(\beta t)= \frac{\delta(t)}{\beta}$.

The similarity representation is given by

$$\frac{d^2F}{d\varsigma^2} + 2\varsigma\frac{dF}{d\varsigma} + 2F = 0 \tag{6.29}$$

$$F(\varsigma\rightarrow\pm\infty) = 0 \tag{6.30a}$$

and

$$\int_{-\infty}^{+\infty} F(\varsigma)d\varsigma = \frac{1}{2} \tag{6.30b}$$

Eqs.(6.29) and (6.30) can be integrated to give the following expression for Green's function:

$$g(\bar{x},\bar{t}) = \frac{Q}{\rho c}\frac{H(\bar{t})}{2\sqrt{\pi\lambda\bar{t}}}exp(\frac{\bar{x}^2}{4\lambda\bar{t}}) \tag{6.31}$$

where $H(\bar{t})$ is the Heaviside function.

In terms of the original variables,

$$g(x,t;x_0,t_0) = \frac{Q}{\rho c}\frac{H(t-t_0)}{2\sqrt{\lambda\pi(t-t_0)}}exp(-\frac{(x-x_0)^2}{4\lambda(t-t_0)}) \tag{6.32}$$

which is a well-known result that can be obtained by integral transforms.

Example 6.4: The Euler-Poisson-Darboux Equation

The equation for the problem of wave propagation in elastic rods of variable cross-section [5] and isentropic flow for a polytropic gas [6] can be written as:

$$L(u) = u_{xx} + \frac{\lambda}{x}u_x - u_{tt} = 0 \tag{6.33a}$$

The fundamental solution to Eq.(6.33a) is the solution of

$$L(u) = \delta(x-x_0)\delta(t-t_0) \tag{6.33b}$$

We will use the infinitesimal approach of Bluman and Cole to find invariant solutions to Eq.(6.33b) [7].

An infinitesimal group of transformation is defined as follows:

$$
\begin{aligned}
\bar{u} &= u + \epsilon f(x,t,u) + O(\epsilon^2) \\
\bar{x} &= x + \epsilon X(x,t) + O(\epsilon^2) \\
\bar{t} &= t + \epsilon T(x,t) + O(\epsilon^2)
\end{aligned}
\tag{6.34}
$$

f, X and T are the infinitesimals to be determined through invariance of Eq.(6.33b), which can be written as

$$
\begin{aligned}
&\bar{u}_{\bar{x}\bar{x}} - \bar{u}_{\bar{t}\bar{t}} + \frac{\lambda}{\bar{x}}\bar{u}_{\bar{x}} - \delta(\bar{x}-x_0)\delta(\bar{t}-t_0) \\
&= u_{xx} - u_{tt} + \frac{\lambda}{x}u_x - \delta(x-x_0)\delta(t-t_0) \\
&+\epsilon[2u_{tx}(X_t - T_x) + u_t(T_{tt} - 2f_t - T_{xx} - \frac{\lambda}{x}T_x) \\
&+u(f_{xx} - f_{tt} + \frac{\lambda}{x}f_x) + u_{xx}(f - 2X_x) + u_{tt}(2T_t - f) \\
&+u_x(2f_x - X_{xx} + X_{tt} - \frac{\lambda}{x}X_x - \frac{\lambda}{x}X + \frac{\lambda}{x}f) \\
&+(X_x - T_t)\delta(x-x_0)\delta(t-t_0)]
\end{aligned}
\tag{6.35a}
$$

with

$$
X(x_0, t_0) = T(x_0, t_0) = 0 \tag{6.35b}
$$

Readers should note that

$$
\delta(\bar{x}) = \delta(x) + \epsilon X(x,0)\delta'(x) + O(\epsilon^2)
$$

Substituting the equation

$$
u_{xx} = -\frac{\lambda}{x}u_x + u_{tt} + \delta(x-x_0)\delta(t-t_0)
$$

in Eq.(6.35), and equating to zero the resulting coefficients of derivatives of u, the following determining equations can be obtained:

$$
T_x - X_t = 0 \tag{6.36a}
$$

$$
T_{tt} - T_{xx} - 2f_t - \frac{\lambda}{x}T_x = 0 \tag{6.36b}
$$

$$
f_{xx} - f_{tt} + \frac{\lambda}{x}T_x = 0 \tag{6.36c}
$$

$$T_t - X_x = 0 \tag{6.36d}$$

$$2f_x - X_{xx} + X_{tt} + \frac{\lambda}{x}X_x - \frac{\lambda}{x^2}X = 0 \tag{6.36e}$$

$$f(x_0, t_0) + T_t(x_0, t_0) - X_x(x_0, t_0) = 0 \tag{6.36f}$$

From Eqs.(6.36d) and (6.36f), we have

$$f(x_0, t_0) = 0 \tag{6.37}$$

Substituting Eq.(6.36a) into Eq.(6.36b) and substituting Eq.(6.36d)

$$2f + \frac{\lambda}{x}X = A(x) \tag{6.38}$$

where A(x) is arbitrary.

Eqs.(6.36a) and (6.36d) give

$$X_{xx} - X_{tt} = 0 \tag{6.39}$$

Thus,solving for f in Eq.(6.36e), we find that A(x)=α=arbitrary constant. However, on account of Eq.(6.35b) and Eq.(6.38), α equals to zero. Therefore,

$$X = -\frac{2x}{\lambda}f \tag{6.40}$$

Next, we find that

$$X_{xx} - X_{tt} = 2\frac{f_x}{\lambda} + \frac{xf_{xx}}{\lambda} - \frac{xf_{tt}}{\lambda} = 0 \tag{6.41}$$

Comparing this with Eq.(6.36c)

$$f_x = 0 \quad ; \quad f_{tt} = 0 \tag{6.42}$$

Therefore, using Eq.(6.37), we get

$$f = \beta(t - t_0) \tag{6.43}$$

and using Eq.(6.40)

$$X = -\frac{2x\beta}{\lambda}(t - t_0) \tag{6.44}$$

From Eqs.(6.36a) and (6.36d),

$$T = \frac{\beta}{\lambda}[2tt_0 - t^2 - x^2] + \gamma \tag{6.45}$$

Using Eq.(6.36b) and setting β equals to 1, we get:

$$f = t - t_0 \quad ; \quad X = -\frac{2x}{\lambda}(t - t_0) \tag{6.46a, b}$$

$$T = \frac{(x_0^2 - x^2) - (t - t_0)^2}{\lambda} \tag{6.46c}$$

The similarity variable is given by

$$u(x,t) = x^{-\lambda/2} g(\varsigma) \quad where \quad \varsigma = x - \frac{(t-t_0)^2}{x} + \frac{x_0^2}{x} \tag{6.47}$$

with $u(x_0,t_0)=1$.

The ordinary differential equation is

$$g''(\varsigma^2 - 4x_0^2) + 2\varsigma g' + \frac{\lambda}{2}(1 - \frac{\lambda}{2})g = 0 \tag{6.48}$$

By letting

$$\phi = \frac{\varsigma + 2x_0}{4x_0},$$

Eq.(6.48) can be written in the hypergeometric form:

$$\phi(1-\phi)\frac{d^2g}{d\phi^2} + (1-2\phi)\frac{dg}{d\phi} - \frac{\lambda}{2}(1 - \frac{\lambda}{2})g = 0 \tag{6.49}$$

The solution can be written as

$$g(\phi) = A_0 F(\frac{\lambda}{2}, 1 - \frac{\lambda}{2}, 1,; \phi) \tag{6.50}$$

where A_0 is some arbitrary constant.

Using the relations,

$$F(a,b;c;z) = (1-z)^{-a} F(a; c-b; c; \frac{z}{z-1})$$

and

$$F(a,b;c;z) = \frac{\Gamma(c)\Gamma(b-a)}{\Gamma(b)\Gamma(c-a)}(-z)^{-a} F(a, 1-c+a; 1-b+a; z^{-1})$$

$$+(-z)^{-b}\frac{\Gamma(c)\Gamma(a-b)}{\Gamma(a)\Gamma(c-b)} F(b, 1-c+b; 1-a+b; z^{-1}) \tag{6.51}$$

Thus we have

$$u = B(x\phi)^{-\lambda/2} F(\frac{\lambda}{2}, \frac{\lambda}{2}, 1, \frac{\phi - 1}{\phi}) \quad with \quad Bx_0^{-\lambda/2} = 1 \tag{6.52}$$

In terms of the original variables

$$u(x,t) = \frac{(2x_0)^\lambda}{[(x+x_0)^2 - (t-t_0)^2]^{\lambda/2}} F\left(\frac{\lambda}{2}, \frac{\lambda}{2};\right.$$

$$\left. 1; \frac{(x-x_0)^2 - (t-t_0)^2}{(x+x_0)^2 - (t-t_0)^2}\right) \tag{6.53}$$

which is the required fundamental solution.

6.3 Pseudo-Similarity Transformations

For a boundary value problem to have a similarity solution, the governing differential equations and the auxiliary conditions should be invariant under a group of transformations. However, if any of the auxiliary conditions is not invariant under the group, then a non-similar description would result. We will discuss a technique which utilizes the similarity variables to obtain non-similar solutions for boundary value problems. The technique makes use of what we will call the "pseudo-similarity transformation" to obtain the required non-similar solution.

Example 6.5 Ground Water Movement due to Arbitrary Change in Water Level[8].

The equation for the flow of ground water which uses the Dupuit-Forchheimer idealization[9] can be written as

$$K\frac{\partial}{\partial x}\left(h\frac{\partial h}{\partial x}\right) = V\frac{\partial h}{\partial t} \tag{6.54}$$

where K is the permeability of the homogeneous and isotropic aquifer, V is the void ratio, h is the height of the water table above the impermeable surface and t is time. It is assumed that all the flow takes place below the water table and the aquifer rests upon an impermeable horizontal bed (Figure 6.4). The auxiliary conditions for the problem can be written as

$$h(x,0) = h_0 \;\; ; \;\; h(x \to \infty; t) = h_0 \;\; and \;\; h(0,t) = H(t) \tag{6.55a, b, c}$$

where $H(t)$ is an arbitrary variation of water level at x=0.

We will non-dimensionalize the equations by introducing the following variables:

$$\bar{x} = \frac{x}{L}\,, \; \bar{t} = \frac{Kh_0 t}{L^2}\,, \; \beta(t) = \frac{H(t)}{h_0}\,, \; \bar{h}(\bar{x},\bar{t}) = \frac{h(x,t)}{h_0}$$

where L is the characteristic length.

Eqs.(6.54) and (6.55) then become

$$\frac{\partial}{\partial \bar{x}}\left(\bar{h}\frac{\partial \bar{h}}{\partial \bar{x}}\right) = \frac{\partial \bar{h}}{\partial \bar{t}} \tag{6.56}$$

and

$$\bar{h}(\bar{x},0) = 1 \;\; ; \;\; \bar{h}(\infty,\bar{t}) = 1 \;\; ; \;\; \bar{h}(0,\bar{t}) = \beta(\bar{t}) \tag{6.57a, b, c}$$

It can be easily seen that Eq.(6.57c) would prevent invariance of the above equations under a one-parameter group of transformations that would have led to a similarity solution. Therefore, the problem description is non-similar. The "pseudo-similarity transformation" can be obtained by ignoring the source of non-similarity, which for the present problem is Eq.(6.57c).

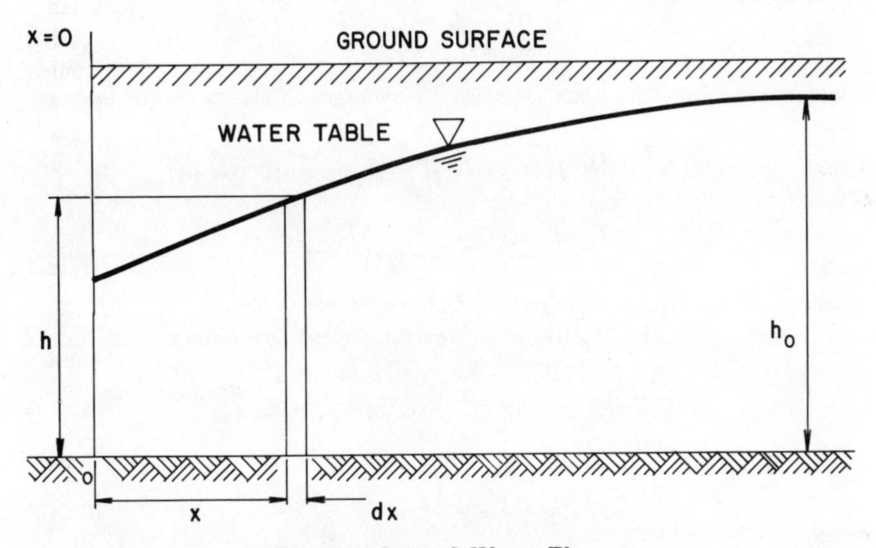

Fig. 6.4 Ground Water Flow

For the remaining problem description, the similarity transformation would be:

$$\varsigma = \frac{\bar{x}}{\sqrt{\bar{t}}} \;\; ; \;\; f(\varsigma) = \bar{h}(\bar{x},\bar{t}) \tag{6.58}$$

For the non-similar problem, the pseudo-similarity transformation can be written as

$$\tau = \bar{t} \; ; \; \varsigma = \frac{\bar{x}}{\sqrt{\bar{t}}} \; ; \; f(\varsigma,\tau) = \bar{h}(\bar{x},\bar{t}) \tag{6.59}$$

In other words, the number of independent variables have not been reduced.

Eqs.(6.56) and (6.57) are now transformed to the following form:

$$ff'' + (f')^2 + \frac{1}{2}\varsigma f' = \tau\frac{\partial f}{\partial \tau} \tag{6.60}$$

subject to the boundary conditions:

$$f(0,\tau) = \beta(\tau) \;\; ; \;\; f(\infty,\tau) = f(\infty,0) = 1 \qquad (6.61a,b,c)$$

where the primes in Eq.(6.60) represent differentiation with respect to ς.

There are some advantages to using the above form, Eq.(6.60). The primary advantage is that the starting process is very simple. At τ=0 all dependency on τ is removed, leaving only an ordinary differential equation to be solved. Often, this equation is a similarity representation for which the solutions are known. For τ greater than 0, non-similar solutions can be obtained by applying "successive corrections" to the similarity solutions.

Eqs.(6.60) and (6.61) can now be solved numerically as follows: As a first step, the derivative in the τ-direction is replaced by a finite difference approximation

$$\frac{\partial f}{\partial \tau} = \frac{f_n - f_{n-1}}{\Delta \tau} \qquad (6.62)$$

The function f and its ς are replaced by averages in the following manner:

$$\frac{1}{2}[(f_n f_n'' + (f_n')^2 + \frac{1}{2}\varsigma(f_n')) + (f_{n-1}f_{n-1}'' + (f_{n-1}')^2$$
$$+\frac{1}{2}\varsigma(f_{n-1}'))] = \tau_{n-1/2}(\frac{f_n - f_{n-1}}{\Delta\tau}) \qquad (6.63)$$

where f_n refers to f at time τ, and f_{n-1} to time $\tau - \Delta\tau$.

Rewriting Eq.(6.63) after some rearrangement, we obtain

$$f_n f_n'' + (f_n')^2 + \frac{1}{2}\varsigma f_n' - \alpha_{n-1/2} f_n = R_{n-1} \qquad (6.64a)$$

where

$$\alpha_{n-1/2} = \frac{2\tau_{n-1/2}}{\Delta\tau} = \frac{\tau_n + \tau_{n-1}}{\Delta\tau} \qquad (6.64b)$$

and

$$R_{n-1} = -\alpha_{n-1/2} f_{n-1} - f_{n-1} f_{n-1}'' - (f_{n-1}')^2 - \frac{1}{2}\varsigma(f_{n-1}') \qquad (6.64c)$$

The right hand side of Eq.(6.64a), which is a recursion scheme, is known. By letting n=0,1,2,.....,etc., in Eq.(6.64a),a sequence of equations for the solution of f_0, f_1, f_2,.......,etc.,and their derivatives can be obtained.

The boundary conditions can correspondingly be written as

$$f_n(\tau,0) = \beta(\tau) \;\; ; \;\; f_n(\tau,\infty) = 1 \qquad (6.65)$$

The solution of Eq.(6.60) subject to boundary conditions, Eq.(6.61), can be obtained by the series

$$f(\tau,\varsigma) = f_0(\varsigma) + \tau f_1(\varsigma) + \tau^2 f_2(\varsigma) + \ldots\ldots \qquad (6.66)$$

where the first term on the right-side represents the similar term and the rest of the series show deviation from similarity. Alternatively, Eqs.(6.64) and (6.65) can be solved by numerical methods such as quasilinearization [10].

Consider the variations of $\beta(\tau)$,

$$(a) \qquad \beta(\tau) = 1 + b\, sin(\frac{\tau}{\tau_0}\frac{\pi}{2}) \qquad \tau \leq \tau_0 \tag{6.67a}$$

$$= 1 + b \qquad \tau > \tau_0 \tag{6.67b}$$

$$(b) \qquad \beta(\tau) = 1 + b\, sin(\frac{\tau}{\tau_0}\frac{\pi}{2}) \qquad \tau \geq 0 \tag{6.68}$$

At τ=0, Eq.(6.60) becomes

$$f_0 f_0'' + (f_0')^2 + \frac{1}{2}\varsigma f_0' = 0 \tag{6.69}$$

An inspection of the initial conditions shows that the solution is

$$f_0(0,\varsigma) = 1 \quad ; \quad f_0' = f_0'' = 0$$

Setting n=1 in Eq.(6.64), the recursion scheme can be initiated. The numerical solution is plotted in Figure 6.5.

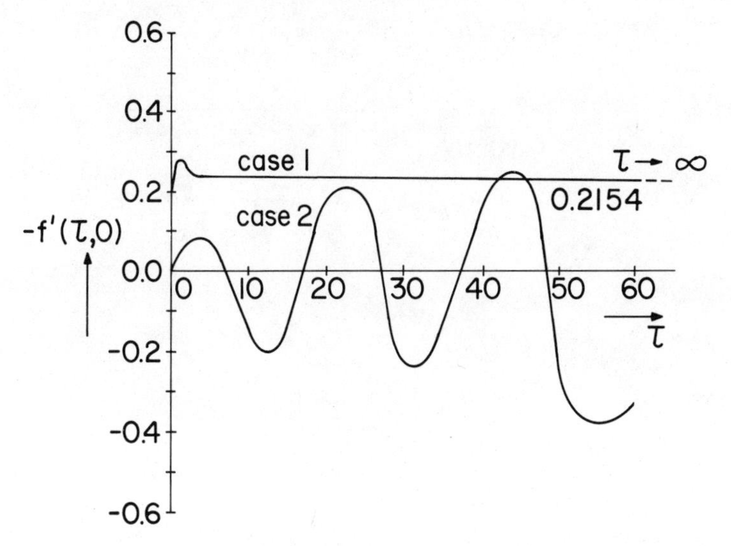

Fig. 6.5 Plot of f'(τ,0) vs τ

Example 6.6 Natural Convection Over Non-isothermal Plate

The non-similar problem of natural convection boundary layer flow over a semi-infinite flat plate (see Figure 6.6) is considered as a second example to illustrate the use of the pseudo-similarity transformation.

The governing differential equations for the solution of natural convection flow past a semi-infinite vertical flat plate with an arbitrary wall temperature, T_w(x), can be written in a non-dimensional form as

$$\frac{\partial \bar{u}}{\partial \bar{x}} + \frac{\partial \bar{v}}{\partial \bar{y}} = 0 \ \ (continuity) \tag{6.70a}$$

$$\bar{u}\frac{\partial \bar{u}}{\partial \bar{x}} + \bar{v}\frac{\partial \bar{u}}{\partial \bar{y}} = \frac{\partial^2 \bar{u}}{\partial \bar{y}^2} + S_w(\bar{x})\theta \ \ (momentum) \tag{6.70b}$$

$$\bar{u}\Big[\frac{\partial \theta}{\partial \bar{x}} + \theta\frac{dln[S_w(\bar{x})]}{d\bar{x}}\Big] + \bar{v}\frac{\partial \theta}{\partial \bar{y}} = \frac{1}{Pr}\frac{\partial^2 \theta}{\partial \bar{y}^2} \ \ (energy) \tag{6.70c}$$

The boundary conditons are

$$\bar{y} = 0 \ : \ \bar{u} = 0 \ , \ \bar{v} = \bar{v}_w(x) \ ; \ \theta = 1 \tag{6.71a}$$

$$\bar{y} \to \infty \ : \ \bar{u} = 0 \ , \ \theta = 0 \tag{6.71b}$$

The dimensionless quantities in Eqs.(6.70) and (6.71) are related to their corresponding physical variables through the following relationships:

$$\bar{x} = \frac{x}{L} \ , \ \bar{y} = \frac{y}{L}\sqrt{Re} \ , \ \bar{u} = \frac{u}{u_c} \ , \ \bar{v} = \frac{v}{u_c} \ , \ \bar{v} = \frac{v}{u_c}\sqrt{Re}$$

$$\theta = \frac{T - T_\infty}{T_w(x) - T_\infty} \ , \ u_c = \sqrt{g\beta(T_r - T_\infty)L}$$

$$Re = \frac{u_c L}{\nu} \ \ and \ \ S_w(\bar{x}) = \frac{T_w(x) - T_\infty}{T_r - T_\infty}$$

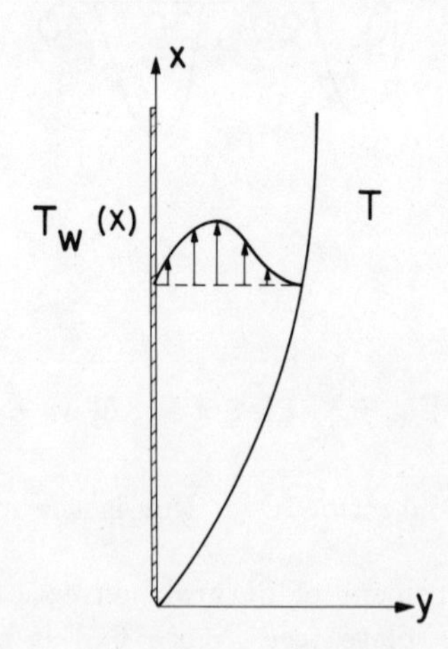

Fig. 6.6 Schematic Sketch of the Plate

L and T_r are reference length and temperature, respectively; x and y are the coordinates parallel and perpendicular to the plate, respectively; Re is the Reynolds number; β is the bulk modulus and subscripts r and ∞ refer to the reference and mainstream conditions, respectively.

Next, the following pseudo-similarity transformation will be introduced for the non-similar problem description:

$$\xi = \bar{x} \; ; \; \eta = \frac{\bar{y}}{\bar{x}^{1/4}[S_w(\bar{x})]^{-1/4}}$$

$$f(\xi,\eta) = \frac{\psi}{\bar{x}^{3/4}S_w(\bar{x})^{1/4}} \tag{6.72}$$

$$g(\xi,\eta) = \theta$$

where the stream function is defined by:

$$\bar{u} = \frac{\partial\psi}{\partial\bar{y}} \quad and \quad \bar{v} = -\frac{\partial\psi}{\partial\bar{x}} \tag{6.73}$$

Eqs.(6.70) then transform into

$$f''' + (\frac{3+P(\xi)}{4})ff'' - (\frac{1+P(\xi)}{2})(f')^2 + g$$

$$= \xi(f'\frac{\partial f'}{\partial\xi} - f''\frac{\partial f}{\partial\xi}) \tag{6.74a}$$

$$\frac{1}{Pr}g'' + \frac{3+P(\xi)}{4}fg' - P(\xi)f'g$$

$$= \xi(f'\frac{\partial g}{\partial\xi} - g'\frac{\partial f}{\partial\xi}) \tag{6.74b}$$

The boundary conditions are transformed to:

$$\eta = 0 \; : \; f'(\xi,0) = 0 \; ; \; g(\xi,0) = 1$$

$$\frac{df}{d\xi} + [\frac{3+P(\xi)}{4\xi}]f = -\frac{M(\xi)}{\xi} \tag{6.75}$$

$$\eta \rightarrow \infty \; : \; f'(\xi,\infty) = 0 \; ; \; g(\xi,\infty) = 0$$

where the temperature and the mass transfer functions are defined as

$$P(\xi) = \frac{\xi}{S_w(\xi)}[\frac{dS_w(\xi)}{d\xi}] \tag{6.76a}$$

and

$$M(\xi) = \bar{v_w}[\frac{\xi}{S_w(\xi)}]^{1/4} \tag{6.76b}$$

The primes denote differentiation with respect to η. Eqs.(6.74) are in a general form. They are for laminar natural convection flows over a semi-infinite plate with arbitrary wall temperature and surface mass transfer. When $P(\xi)$ and $M(\xi)$ are constants,i.e.,

$$\frac{\xi}{S_w(\xi)}\frac{dS_w(\xi)}{d\xi} = \alpha \quad and \quad \bar{v_w}\left[\frac{\xi}{S_w(\xi)}\right]^{1/4} = \beta$$

the wall temperature and surface mass transfer distributions are given by:

$$S_w(\xi) = c_1\xi^{\alpha} \quad ; \quad \bar{v_w}(\xi) = c_2\xi^{(\alpha-1)/4} \tag{6.77}$$

The problem description now becomes similar and the right-hand sides of Eq.(6.74) becomes zero, since for this case, both f and g are independent of ξ.

Again, the solution can be obtained by expressing f and g as follows:

$$f(\eta,\xi) = f_0(\eta) + \xi f_1(\eta) + \xi^2 f_2(\eta) + ...$$

$$g(\eta,\xi) = g_0(\eta) + \xi g_1(\eta) + \xi^2 g_2(\eta) + ...$$

Alternately, Eqs.(6.74) and (6.75) can be solved by numerical procedures.

Example 6.7 Pseudoplastic Non-Newtonian Flow Near a Moving Plate [12]

The Ostwald-de-Waele model for pseudoplastic non-Newtonian flow can be written as

$$\tau_{xy} = -m\left|\frac{\partial v_x}{\partial y}\right|^{n-1}\frac{\partial v_x}{\partial y} \tag{6.78}$$

where τ_{xy} is the shear stress, v_x is the velocity along the x axis at a distance of y from the wall surface, m and n are constants of the pseudoplastic fluid.

The equation of motion for the system is given by

$$\rho\frac{\partial v_x}{\partial t} = -\frac{\partial \tau_{xy}}{\partial y} \tag{6.79}$$

where ρ is the density of the fluid. Combining Eqs.(6.78) and (6.79), the nonlinear partial differential equation for the velocity distribution can be written as

$$\rho\frac{\partial v_x}{\partial t} = c\ m\ \frac{\partial}{\partial y}\left(c\frac{\partial v_x}{\partial y}\right)^n \tag{6.80}$$

where

$$c = -1 \quad when \quad \frac{\partial v_x}{\partial y} < 0$$

$$c = 1 \quad when \quad \frac{\partial v_x}{\partial y} > 0$$

The boundary conditions for the problem are given by

$$v_x(0,t) = v_0\phi(t) \;\; ; \;\; v_x(\infty,t) = 0 \tag{6.81}$$

The initial condition is given by

$$v_x(y,0) = 0 \tag{6.81c}$$

The prescription of an arbitrary $\phi(t)$ in Eq.(6.81c) would prevent invariance of the problem as described in Eqs.(6.80) and (6.81), and therefore, the solution would be non-similar.

We now introduce the non-dimensional variables

$$\bar{v}_x = \frac{v_x}{v_0} \; ; \; \bar{t} = (\frac{\rho v_0^2}{m})^{1/n} t \; ; \; \bar{y} = (\frac{\rho}{m v_0^{n-2}})^{1/n} y$$

The pseudo-similarity transformation can be written as

$$\bar{v}_x = \phi(\tau) f(\tau, \varsigma) \tag{6.82}$$

where

$$\tau = \bar{t} \; ; \; \varsigma = \frac{\bar{y}}{\bar{t}^{\,1/(n+1)}}$$

Eqs.(6.80) and (6.81) become:

$$c\phi^{n-1}[(cf')^n]' + \frac{1}{n+1}\varsigma f' - P(\tau) f = \tau \frac{\partial f}{\partial \tau} \tag{6.83}$$

where

$$P(\tau) = \frac{\tau}{\phi}(\frac{d\phi}{d\tau}) \tag{6.83}$$

and the boundary conditions are:

$$f(0,\tau) = 1 \;\; ; \;\; f(\infty,\tau) = 0 \tag{6.84a, b}$$

When τ=0, the solution is similar. For τ greater than zero, solutions can be obtained as described earlier in this chapter.

6.4 Similarity Solutions as Asymptotic Limits of the Non-Similar Problem Description

Similarity solutions resulting from invariance under dimensional and affine groups of transformations are quite often of interest, because they are limits that are asymptotically approached by solutions of the more general problems that are non-similar. The closer the initial condition is to the "limiting solution", the earlier will the non-similar problem approach the similar regime.

The basic nature of the asymptotic behavior of a solution can be best illustrated by using the example of linear heat conduction again, for which the initial temperature distribution $T(x,0)=f(x)$ along the x axis and the boundary description $T(\sigma,t)=h(t)$ is given. The problem statement is as follows:

$$(\frac{\partial}{\partial t} - \frac{\partial^2}{\partial x^2})\, T(x,t) = q(x,t) \tag{6.85}$$

$$(-\infty < x < \infty \; ; \; t > 0)$$

$$T(x,0) = f(x) \quad ; \quad T(\sigma,0) = h(t) \tag{6.86a, b}$$

$h(t)$ is the temperature prescribed at the boundary surface, $x=\sigma$.

Using similarity techniques, the Green's function $g(x,t\,|\,x_0, t_0)$ can be determined. The general solution to the overall problem can be written as[4]:

$$T(x,t) = \int_0^t \int_{-\infty}^{\infty} q(x_0, t_0)\, g(x,t|x_0,t_0)\, dx_0\, dt_0$$

$$+ \int_{-\infty}^{\infty} f(x_0)\, g(x,t|x_0,t_0)\, dx_0$$

$$- \int_0^t dt_0 \int_\sigma \frac{\partial g(x,t|x_0,t_0)}{\partial n} h(t_0)\, dS_0 \tag{6.87}$$

We will further stipulate that $h(t)=0$. If

$$q(x,t) = \delta(x - x_0)\delta(t),$$

the Green's function $g(x,t|x_0,0)$ can be written as [see Eq.(6.32)]

$$g(x,t|x_0,0) = \frac{H(t)}{\sqrt{4\pi t}} e^{-(x-x_0)^2/4t} \tag{6.88}$$

$H(t)$ is the Heaviside function. Eq.(6.87) can therefore be written as

$$T(x,t) = \frac{H(t)}{\sqrt{4\pi t}} \int_{-\infty}^{\infty} f(x_0)\, exp(-\frac{(x-x_0)^2}{4t})\, dx_0 \tag{6.89}$$

We will now consider the behavior of the temperature as $t \to \infty$. Expanding the integrand, Eq.(6.89) in a power series:

$$T(x,t) = \frac{H(t)}{\sqrt{4\pi t}} \int_{-\infty}^{\infty} (f(x_0)\, exp(-\frac{x^2}{4t} + \frac{2xx_0}{4t} - \frac{x_0^2}{4t})\, dx_0$$

$$= \frac{H(t)}{\sqrt{4\pi t}} \int_{-\infty}^{\infty} e^{-\varsigma^2} f(x_0)[(1 + \frac{xx_0}{2t} + \frac{4x^2x_0^2}{2(4t)^2})$$

$$\cdot(1 - \frac{x_0^2}{4t})]dx_0 \tag{6.90}$$

Simplifying Eq.(6.90), the following expression can be obtained

$$T(x,t) = \frac{H(t)}{\sqrt{4\pi t}} e^{-\varsigma^2} [\int_{-\infty}^{\infty} f(x_0)\,dx + \frac{\varsigma}{\sqrt{t}} \int_{-\infty}^{\infty} x_0 f(x_0)\,dx_0$$

$$+ \frac{2\varsigma^2 - 1}{4t} \int_{-\infty}^{\infty} x_0^2 f(x_0)\,dx_0 + ...] + ... \tag{6.91}$$

where $\varsigma = x/(2\sqrt{t})$ is the similarity variable.

Eq.(6.91) can be seen to be a sum of similarity terms in which the powers of the inverse time increase by 1/2 with each successive term, and the coefficients are expressed in terms of successive moments of the initial temperature distribution. In the limit $t \rightarrow \infty$, only the first term of Eq.(6.91) remains. This corresponds to the concentrated source solution as expressed by Eq.(6.88). The subsequent terms in the expansion characterize the difference between the actual and limiting solution.

Eq.(6.91) can be expressed in a different form as follows:

$$T(x,t) = T_{lim}(1 + \frac{\psi(\varsigma)}{\sqrt{t}} + ...) \tag{6.92}$$

where

$$T_{lim} = \frac{H(t)}{\sqrt{4\pi t}} e^{-\varsigma^2} [\int_{-\infty}^{\infty} f(x_0)\,dx_0]$$

and

$$\psi(\varsigma) = \frac{\varsigma \int_{-\infty}^{\infty} x_0 f(x_0)\,dx_0}{\int_{-\infty}^{\infty} f(x_0)\,dx_0}$$

The asymptotic validity appears to exist for invariance under dimensional or affine groups of transformations. The resulting similarity solutions are often referred to as "self-similar solutions" [14] (see section 11.5).

6.5 Summary

When invariance of a set of partial differential equations under a group of transformation cannot be invoked either partially or entirely, then a non-similar representation exists. In this chapter, some methods for obtaining non-similar solutions from similarity solutions that have been examined are:

(a) superposition of similarity solutions

(b) fundamental solutions

(c) pseudo-similarity solutions

Similarity solutions obtained by the invocation of invariance under an affine or a dimensional group have been shown to be asymptotic limits of the non-similar problem description.

REFERENCES

1. Hansen, A.G.,Similarity Analysis of Boundary Value Problems in Engineering, Prentice-Hall (1964).
2. Na, T.Y. and Hansen, A.G.,"Similarity Analysis of Flow Near an Oscillating Plate",ASME paper 65-FE-21, presented in the ASME Fluid Engineering Conference, July 12, 1965.
3. Morgenstern,N.R. and Nixon, J.F.,"One-Dimensional Consolidation of Thawing Soils", Can. Geotech. J.,8 (1971).
4. Stakgold, I., Boundary Value Problems of Math. Phys., Vol.II, MacMilan, 1968.
5. Cristescu, N., Dynamic Plasticity, North Holland Pub. Co., Amsterdam (1967).
6. Garabedian, P.R.,Partial Differential Equations, Wiley, (1964).
7. Bluman, G.W., Construction of Solutions to Partial Differntial Equations by the use of Trnsformation Groups, Ph.D. Thesis, California Institute of Technology (1967).
8. Seshadri, R. and Na, T.Y.,"Ground Water Movement Due to Arbitrary Changes in Water Level", Appl. Sci. Res.,39 (1982).
9. Bear, J., Dynamics of Fluids in Porous Media, American Elsevier Pub. Co. (1972).
10. Na, T.Y., Computational Methods in Engineering Boundary Value Problems, Academic Press Inc. (1979).
11. Na, T.Y., "Numerical Solution of Natural Convection Flow Past a Non-Isothermal Vertical Flat Plate", Appl. Sci. Res., 33 (1978).
12. Na, T.Y., Seshadri, R. and Singh, M.C., "On Obtaining Non-similar Solutions from Similarity Solutions", 4th Int'l Sym. On Large Eng. Systems, University of Calgary, Calgary, Canada (1982).
13. Bird, R.B.,"Unsteady Pseudo-plastic Flow Near a Wall", A.I.Ch.E. Journal, Vol.5, No.4 (1959).
14. Zeldovich, Ya. B. and Raizer, Yu. P., Physics of Shock Waves and High Temperature Hydrodynamic Phenomena, (Translation) Vol.2, Academic Press (1967).

Chapter 7

MOVING BOUNDARY PROBLEMS GOVERNED BY PARABOLIC EQUATIONS

7.0 Introduction

In order to carry out similarity analysis of moving boundary problems governed by parabolic partial differential equations, it is necessary to ascertain whether the speed of propagation of the moving boundary is infinite or finite. This can usually be accomplished by carefully investigating the physical formulation of the boundary value problem. For the purpose of similarity analysis, such problems can be classified into (1) problems that involve a change of phase, and (2) problems without a change of phase. It will be seen that in boundary value problems with a change of phase, the moving boundary would advance with a finite speed of propagation. However, the moving boundary could propagate with either infinite or finite speed in problems where no phase change is involved.

The classification of a second order quasilinear partial differential equation will be briefly reviewed. Consider a boundary value problem that is described by an equation with a dependent variable, u, and independent variables, x and t, such that

$$Au_{tt} + Bu_{xt} + Cu_{xx} + D = 0 \tag{7.1}$$

where A, B, C and D are functions of x, t, u, u_x and u_t. Eq.(7.1) is called quasilinear because it is linear in the derivative of the highest order. The notion of characteristics could be introduced as the loci of possible small discontinuities. For Eq.(7.1), two families of characteristics can be defined as follows:

$$\left(\frac{dx}{dt}\right)_1 = \frac{B + \sqrt{B^2 - 4AC}}{2A} \tag{7.2a}$$

$$\left(\frac{dx}{dt}\right)_2 = \frac{B - \sqrt{B^2 - 4AC}}{2A} \tag{7.2b}$$

Depending on the values of the functions A, B, C and D, the classification criteria can be described as in the table below:

Eq. Type	$(B^2 - 4AC)$	Char. Curves
Parabolic	$= 0$	1 real family
Hyperbolic	> 0	2 real families
Elliptic	< 0	2 imaginary families

Since A, B and C are, in general, variables which can take on different values in different parts of the region, Eq.(7.1) may be hyperbolic in some parts, parabolic or elliptic in others.

7.1 Problems With Phase Change

The moving boundary for boundary value problems that involve a phase change propagate at a finite speed. The speed of propagation can usually be determined by applying some form of conservation relationship at the moving boundary. In this section, the well known problem of one-dimensional freezing of a liquid is used for the purpose of illustration. The exact solution to this classical problem is referred to as Neumann's solution.

The problem considered is one in which a semi-infinite region is held at an initial constant temperature, T_0. The temperature of the surface is suddenly dropped to T_s, and held constant thereafter(see Figure 7.1). It is assumed that,initially,the medium is in a liquid state, $T_0 > T_f$,where T_0 is the initial temperature and T_s is the surface temperature.

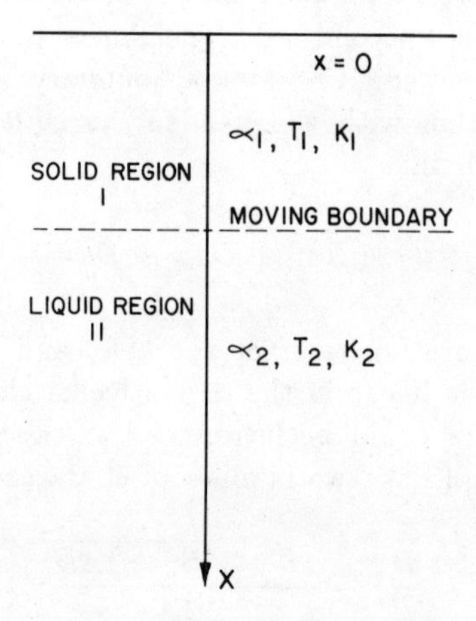

Fig. 7.1 Phase Change Problem

The boundary value problem can be formulated as follows:

$$\alpha_1 \frac{\partial^2 T_1}{\partial x^2} = \frac{\partial T_1}{\partial t} \quad ; \quad \alpha_2 \frac{\partial^2 T_2}{\partial x^2} = \frac{\partial T_2}{\partial t} \tag{7.3}$$

The boundary conditions are

$$T_1(0,t) = T_s \quad ; \quad T_2(\infty, t) = T_0 \tag{7.4}$$

α_1 and α_2 are the thermal diffusivities in the frozen and unfrozen zones. The position of the freezing interface is a function of time, and the temperature at this location for the liquid as well as the solid is equal to the fusion temperature, T_f. At the moving boundary, the heat balance equation can be expressed as follows:

$$K_1 \frac{\partial T_1}{\partial x} - K_2 \frac{\partial T_2}{\partial x} = \rho L \frac{dx}{dt} \tag{7.5a}$$

and

$$T_1(X(t), t) = T_2(X(t), t) = T_f \tag{7.5b}$$

where K_1 and K_2 are the thermal conductivities in the frozen and unfrozen zones,ρ is the mass density and L is the latent heat of fusion(mass basis).

The solution to this problem can be obtained by using the similarity transformation

$$\varsigma_1 = \frac{x}{2\sqrt{\alpha_1 t}} \tag{7.6}$$

For the frozen region, the resulting ordinary differential equation is

$$\frac{d^2 T_1}{d\varsigma_1^2} + 2\varsigma \frac{dT_1}{d\varsigma} = 0 \tag{7.7}$$

The solution for Eq.(7.7), using the condition T_1 (0,t)=T_s is

$$T_1(x, t) = T_s + A\ erf(\varsigma_1) \tag{7.8}$$

Similarly, for the liquid region

$$T_2(x, t) = T_0 - B\ erf(\varsigma_2) \tag{7.9}$$

where erf( ) is the error function.

Equality of temperature of the solid and the liquid at the interface gives

$$A\ erf(\frac{X}{2\sqrt{\alpha_1 t}}) = T_0 - B\ erfc(\frac{X}{2\sqrt{\alpha_2}}) = T_f \tag{7.10}$$

where erfc( )=1-erf( ).

Eq.(7.10) can only be satisfied if

$$\frac{X}{2\sqrt{\alpha_1 t}} = \gamma = constant$$

Therefore, $X(t) = 2\sqrt{\alpha_1 t}$ would describe the movement of the interface provided γ is known.

The constant,γ, can be determined by using the heat balance equation at the interface, which takes the form

$$\frac{K_1/K_2\sqrt{\alpha_1/\alpha_2}(T_0 - T_f)exp(-\alpha_1/\alpha_2)\gamma^2}{(T_f - T_s)erfc(\gamma\sqrt{\alpha_1/\alpha_2})}$$

$$= \frac{e^{-\gamma^2}}{erf(\gamma)} - \frac{L\gamma\sqrt{\pi}}{c_1(T_f - T_s)} \tag{7.11}$$

where c_1 is the specific heat of ice.

The original moving boundary description becomes a fixed boundary description in the similarity coordinate. The rate of movement of the interface varies with time, i.e.,

$$propagation\ speed = \frac{dX(t)}{dt} = \frac{\gamma\sqrt{\alpha_1}}{\sqrt{t}} \tag{7.12}$$

For any time, $t > 0$, the speed of propagation of the interface is finite.

The term "change of phase" is used to describe not only freezing of liquids, but also to include any physical situation where a moving boundary divides a region under consideration into portions with distinct properties. For example, the problem of impact of viscoplastic rod which has different properties during loading and unloading, would give rise to a moving boundary that separates the loading and unloading regions.

7.2 Problems Without Phase Change

Similarity analysis of parabolic partial differential equations that arise from physical formulations that do not involve a change of phase, requires careful consideration with regard to the location of the moving boundary in the similarity coordinate. Two distinct possibilities should be considered in such cases:

(1) existence of a sharp moving boundary which propagates with a finite speed, and

(2) instantaneous propagation of the moving boundary.

For the purpose of illustration, consider the nonlinear heat equation of the form

$$\frac{\partial}{\partial x}[K(T)\frac{\partial T}{\partial x}] - \frac{\partial T}{\partial t} = Q(x,t) \tag{7.13}$$

where $K(T)$ is the nonlinear property, T is the dependent variable and $Q(x,t)$ is the heat source. In the present discussion, we assume that $Q(x,t) = 0$.

Eq.(7.13) arises in certain diverse physical situation in science and engineering[1]:

(a) the transport of thermal energy by radiation in a completely ionized gas. The coefficient $K(T)$ is equal $K_0 T^n$, where n is equal to 6.5, and in regions of multiply ionized gases n varies between 4.5 and 5.5;

(b) electron heat conduction in plasma, where the coefficient of electron thermal diffusivity $K(T)$ is equal to $\beta T^{5/2}$,

(c) ordinary nonlinear heat conduction phenomena where the thermal conductivity depends on temperature, and

(d) the electric transmission in cables coated with resistive paints that exhibit nonlinear characteristics. T in Eq.(7.13) is replaced by the electric field variable E, and and $K(E) = [E/r(E)]'/c$. c is the capacitance per unit length and $r(E)$ is the resistance per unit length. In certain high voltage applications, conducting cables are sheathed in cylinders of capacitive materials with resistive paints which exhibit nonlinear characteristics. An empirical formula can be written as $r(E) = r_0(E_0/E)^n$. The solution to this problem is one which describes what occurs when a voltage is applied at the end of the transmission line when $t = 0$.

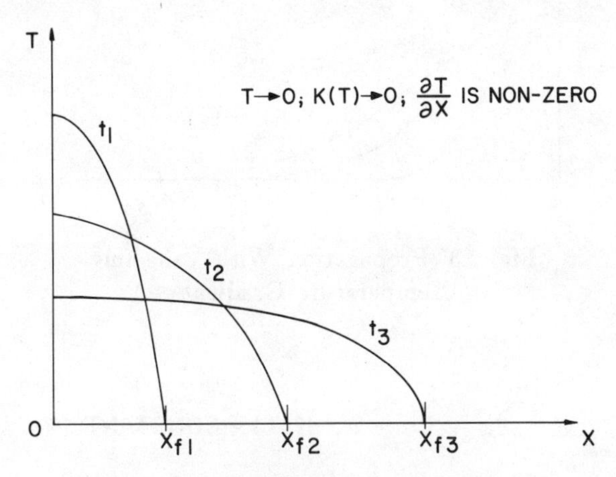

Fig. 7.2 Propagation With a Sharp Moving Boundary

For further discussions, the problem of transport of thermal energy by radiation in a completely ionized gas will be considered. The radiation phenomena comes into effect when temperatures of the order of tens and hundreds of thousands of degrees are encountered. Since $K(T)$ is equal to $K_0 T^n$ in Eq.(7.13), as $T \to 0$, the heat flux $q(x, t) = K(T)(\partial T/\partial x) \to 0$. This implies that $q(x, t) \to 0$ for a nonvanishing gradient, $\partial T/\partial x$. If we assume that a one-dimensional region ahead of the heat source is initially at zero degrees, then a sharp moving boundary can be expected since $\partial T/\partial x$ need not vanish at the moving boundary. With respect to Figure 7.2, X_{f1}, X_{f2} and X_{f3} are successive locations of the moving boundary with the passage of time, for the case of non-vanishing gradient. In some cases, both $K(T)$ and $\partial T/\partial x$ can be simultaneously zero at the moving boundary (see Figure 7.3).

Since $K(T) = K_1$ (constant) as T approaches zero, Eq.(7.13) becomes linear. For gases, K_1 equals to $l_a \bar{v}/3$, where l_a is the molecular mean free path, and $\bar{v}$ is the mean thermal speed. The heat flux $q(x, t)$ will vanish only when the gradient $\partial T/\partial x$ approaches to zero. The decay of temperature with x will be asymptotic and the propagation of the thermal wave can be considered to be instantaneous. The variation of temperature with distance for different times is shown in Figure 7.4.

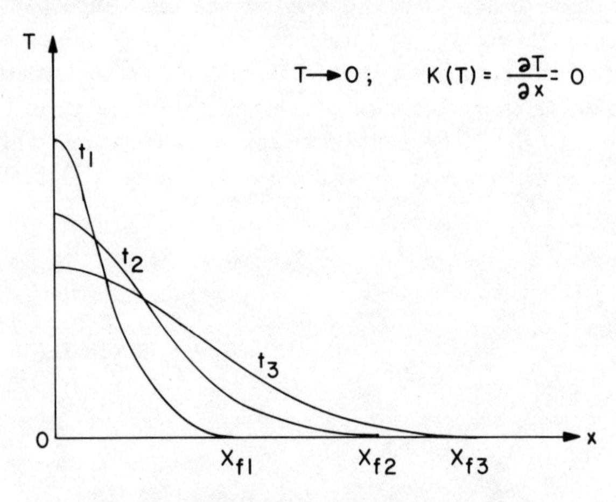

Fig. 7.3 Propagation With Vanishing Temperature Gradient

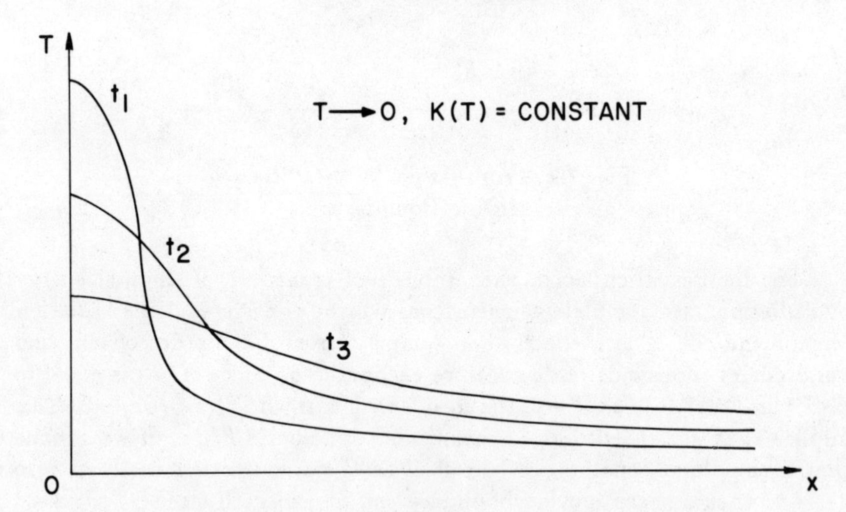

Fig. 7.4 Instantaneous Thermal Propagation

If the heat source is given by

$$Q(x,t) = q_0 \delta(x) \delta(t),$$

then we have

$$T(x,t) = \frac{q_0}{\sqrt{4\pi K_1 t}} \, exp\Big(-\frac{x^2}{4K_1 t}\Big) \tag{7.14}$$

It is also seen that if

$$K(T) = K_1 + K_0 T^n$$

then

$$as \;\; T \rightarrow 0 \;\; , \;\; K(T) \rightarrow K_1$$

In a situation such as this, the propagation would be instantaneous since $\partial T / \partial x$ equals to zero for vanishing flux.

The existence of a sharp moving boundary can usually be determined by studying the moving boundary condition, along with the overall physical basis for the problem.

Example 7.1 : Thermal Waves from an Instantaneous Plane Source[2]

This example further clarifies foregoing discussions on aspects relating to the propagation of sharp moving boundaries in problems that do not involve a phase-change. The propagation of heat from an instantaneous plane source in an initially cool infinite domain is considered. Heat propagates in both directions off the plane $x = 0$ where an energy E per unit area of the surface is released.

The nonlinear heat conduction equation is

$$\frac{\partial}{\partial x}(\alpha T^n \frac{\partial T}{\partial x}) = \frac{\partial T}{\partial t} \tag{7.15}$$

The conservation of energy can be expressed as

$$\int_{-\infty}^{\infty} T(x,t)\, dx = Q \tag{7.16}$$

We use the Hellums-Churchill procedure to determine the similarity transformation. Defining

$$T = \frac{T}{T_0} \;\; ; \;\; \bar{x} = \frac{x}{x_0} \;\; ; \;\; \bar{t} = \frac{t}{t_0}$$

we have

$$\pi_{e1} = \frac{\alpha T_0^n t_0}{x_0^2} \;\; ; \;\; \pi_{b1} = \frac{Q}{T_0 x_0}$$

For minimum parametric description, we set

$$\pi_{e1} = \pi_{e2} = 1$$

Therefore,

$$x_0 = (\alpha Q^n)^{\frac{1}{n+2}} t_0^{\frac{1}{n+2}} \;\; ; \;\; T_0 = \frac{Q^{2/(n+2)}}{(\alpha t)^{1/(n+2)}} \tag{7.17}$$

The similarity transformation can be written as

$$T(x,t) = \left(\frac{Q^2}{\alpha t}\right)^{\frac{1}{n+2}} F(\varsigma), \quad where \ \ \varsigma = \frac{x}{(\alpha Q^n t)^{1/(n+2)}} \tag{7.18}$$

Eq.(7.15) can be transformed into the following ordinary differential equation:

$$(n+2)\frac{d}{d\varsigma}[F^n \frac{dF}{d\varsigma}] + \varsigma\frac{dF}{d\varsigma} + F = 0 \tag{7.19}$$

The boundary conditions are

$$F(\infty) = 0 \ \ and \ \ F'(0) = 0 \ (by \ virtue \ \ of \ symmetry)$$

Integrating Eq.(7.19) once,

$$F^n\frac{dF}{d\varsigma} + \frac{\varsigma F}{(n+2)} = c_1 \tag{7.20}$$

Using the boundary condition $f'(0) = 0$, the constant of integration $c_1 = 0$. Integrating Eq.(7.20), we get

$$F(\varsigma) = \left[c_2 - \frac{n\varsigma^2}{2(n+2)}\right]^{\frac{1}{n}} \tag{7.21}$$

We have to allow for the possibility of the existence of a sharp moving boundary, since $K(T) = \alpha T^n$, and the heat flux vanishes with a non-vanishing gradient as $T \rightarrow 0$. Therefore, the condition $f(\infty) = 0$ is replaced by $f(\varsigma_0) = 0$, where ς_0 locates the moving boundary. Using this condition, c_2 is equal to $n\varsigma_0^2/2(n+2)$. Therefore,the solution becomes

$$F(\varsigma) = [\frac{n}{2(n+2)}(\varsigma_0^2 - \varsigma^2)]^{1/n} \quad for \ \varsigma < \varsigma_0 \tag{7.22}$$

and

$$f(\varsigma) = 0 \qquad for \ \varsigma > \varsigma_0 \tag{7.22b}$$

The value of ς_0 can be determined from the conservation of energy expression:

$$\int_{-\varsigma_0}^{+\varsigma_0} F(\varsigma)d\varsigma = 1 \tag{7.23}$$

Substituting Eq.(7.22) into Eq.(7.23) and simplifying,

$$\varsigma_0 = [\frac{(n+2)^{(1+n)}2^{1-n}}{n\pi^{n/2}}]^{\frac{1}{n+2}}[\frac{\Gamma(1/2+1/n)}{\Gamma(1/n)}]^{\frac{n}{n+2}} \tag{7.24}$$

where $\Gamma(\ )$ is the Gamma function.

The motion of the thermal front can be described by setting $\varsigma = \varsigma_0$, such that

$$X(t) = \varsigma_0 (\alpha Q t)^{\frac{1}{n+2}} \tag{7.25}$$

When the exponent n is equal to zero, we have the linear heat equation. Using Eq.(7.24),

$$\alpha_0 = \frac{2}{\sqrt{n}} \to \infty$$

This confirms our earlier discussions in which it had been stated that, for $K(T)$ is a constant, instantaneous propagation would result. The solution for the linear equation is

$$T(x,t) = \frac{Q}{\sqrt{\alpha t}} [F(\varsigma)]_{n \to 0} \tag{7.26a}$$

or

$$T(x,t) = \frac{Q}{\sqrt{4\pi\alpha t}} e^{-x^2/4\alpha t} \tag{7.26b}$$

Example 7.2 : Multidimensional Diffusion Equation[3]

Consider the m-dimensional equation with spherical symmetry for the diffusion process

$$\frac{1}{r^{m-1}} \frac{\partial}{\partial r} \left[r^{m-1} D(c) \frac{\partial c}{\partial r} \right] = \frac{\partial c}{\partial t} \tag{7.27}$$

c is the concentration, and we assume that $D(c) = c^n$. The auxiliary conditions for $t > 0$ are

$$c(0,t) = t^{\beta} \quad ; \quad c(r,0) = 0 \quad ; \quad c(\infty, t) = 0 \tag{7.28}$$

The similarity transformation can be derived by the Birkhoff-Morgan approach using the one-parameter group (G):

$$G : \bar{c} = A^{\alpha_3} c \quad ; \quad \bar{r} = A^{\alpha_2} r \quad ; \quad \bar{t} = A^{\alpha_1} t$$

For invariance,

$$\frac{\alpha_3}{\alpha_1} = \frac{2(\alpha_2/\alpha_1) - 1}{n}$$

Therefore, we get:

$$c(r,t) = t^q F(\varsigma), \quad and \quad \alpha = \frac{r}{t^s} \tag{7.29}$$

In Eq.(7.29),

$$q = \frac{\alpha_3}{\alpha_1} \quad , \quad s = \frac{\alpha_2}{\alpha_1} \quad and \quad \alpha_3 = \beta .$$

Substituting Eq.(7.29) into Eq.(7.27), we get

$$(\varsigma^{m-1} f' f^n)' = q\varsigma^{m-1} f - s\varsigma^m f' \tag{7.30}$$

If $s = 1/(nm+2)$,Eq.(7.30) can be integrated into the following form:

$$s[\varsigma^{m-1} f' f^n]' + (\varsigma^m f)' = 0 \tag{7.31}$$

Integrating again, we have

$$s f' f^n + \varsigma f = K_1 \varsigma^{1-m} \tag{7.32}$$

where K_1 is a constant of integration. The transformed boundary conditions are:

$$f(0) = 1 \quad \textit{and} \quad f(\infty) = 0$$

Here again, the question of whether or not a sharp moving boundary ($\varsigma = \varsigma_0$) would exist, can be resolved through the physics of the problem. Since $D(c) = c^n$, the flux is $c^n c_x$. Assuming zero concentration ahead of the moving boundary, the flux will vanish for either a non-zero or vanishing gradient.

Eq.(7.32) can be simplified by substituting

$$G(\varsigma) = [f(\varsigma)]^{n+1}$$

into the following form:

$$\frac{1}{n+2}(G') = K_1 \varsigma^{1-m} - \frac{\varsigma G^{1/(n+1)}}{(nm+2)} \tag{7.33}$$

The boundary conditions would be

$$G(0) = 1 \quad ; \quad G(\varsigma_0) = 1 \tag{7.34}$$

In order to evaluate $K_1, G'(\varsigma_0)$ also should be specified. If $G'(\varsigma_0) = 0$, then Eq.(7.33) can be integrated as

$$G(\varsigma) = [\frac{n}{2(nm+2)}(\varsigma_0^2 - \varsigma^2)]^{(n+1)/n} \tag{7.35}$$

Applying the condition $G(0) = 1$, we have

$$\varsigma_0 = [\frac{2(nm+2)}{n}]^{1/2} \tag{7.36}$$

The expression for $c(r,t)$ can now be written as

$$c(r,t) = t^{-\frac{m}{nm+2}} [\frac{n}{2(nm+2)}(\varsigma_0^2 - \varsigma^2)]^{\frac{1}{n}} \tag{7.37}$$

There is another important class of problems for which the equations and the boundary conditions are invariant under a group of translations. Such an invariance leads to "uniform propagation regimes" or "traveling wave solutions" in which the moving boundary propagates with a finite speed. As an example, consider the "power-law" diffusion equation[4]

$$(u^n)\frac{\partial^2 u}{\partial x^2} = \frac{\partial u}{\partial t} \tag{7.38}$$

Eq.(7.38) is invariant under a group of translations

$$G : \quad \bar{x} = x + \epsilon\alpha \quad ; \quad \bar{t} = t + \epsilon\beta \quad ; \quad \bar{u} = u$$

where ϵ is the parameter of transformation.

The similarity variable is given by the integrating of the subsystem

$$\frac{dx}{\alpha} = \frac{dt}{\beta} = \frac{du}{0} \tag{7.39}$$

so that

$$u = f(\varsigma), \quad where \ \ \varsigma = x - \lambda t \ \ and \ \ \lambda = \frac{\alpha}{\beta} \tag{7.40}$$

It must be determined what form f should take so that $f(\varsigma)$ is a solution to Eq.(7.38). Upon substitution of Eq.(7.40) into Eq.(7.38), we find that f must satisfy the differential equation

$$(f^n)'' + \lambda f' = 0 \tag{7.41}$$

where prime indicates differentiation with the similarity variable,ς.

Integrating Eq.(7.41) once,

$$(f^n)' + \lambda f = A \tag{7.42}$$

where A is a constant. Integrating Eq.(7.42) for $n > 0$, the following implicit solution can be found[4]:

$$\sum_{j=0}^{n-2} \frac{(A/\lambda)^j u^{n-1-j}}{n-1-j} + \left(\frac{A}{\lambda}\right)^{n-1} ln\left(u - \frac{A}{\lambda}\right) = \frac{\lambda}{n}(t\lambda - x + \beta) \tag{7.43}$$

where A is non-zero and B is another constant. If $A = 0$, the integration generates the explicit form

$$u(x,t) = \left[\frac{\lambda(n-1)}{n}(\lambda t - x + B)\right]^{\frac{1}{n-1}} \tag{7.44}$$

for $n \neq 1$. However,when n is equal to 1,

$$u(x,t) = B \ exp[-\lambda(x - \lambda t)] \tag{7.45}$$

It can therefore be seen that parabolic equations can have a sharp traveling wave front.

More problems of the traveling wave type will be considered in chapter 8.

7.3 Summary

In this chapter, two types of physical problems were identified with regard to similarity analysis of parabolic differential equations:

(a) problems involving a phase change, and

(b) problems without a phase change.

In problems in which there is a change of phase, the moving boundary propagates at a finite speed. Similarity analysis is carried out for each of the phases, and physical compatibility at the moving boundary is taken into account.

In problems without phase change, the propagation can be instantaneous for certain situations. In other situations, however, moving boundaries could exist that propagate with a finite propagation speed. The process of determining the propagation speed, and whether or not a sharp boundary would exist will depend on the physical nature of the problem. The propagation of a traveling wave in uniform regimes can be discovered by invoking invariance under a group of translations.

REFERENCES

1. Boyer, R.H., "On Some Solutions of a Nonlinear Diffusion Equation", J. of Math. and Phys., Vol.41, 41(1962).
2. Zeldovich,Ya.B. and Raizer, Yu.P., "Physics of Shock Waves and High Temperature Hydrodynamic Phenomena", Hayes,W.D. and Probstein, R.F. (Editors), Vol.2, Academic Press (1967).
3. Ames, W.F., "Similarity for the Nonlinear Diffusion Equation",I and EC Fundamentals, Vol.4, No.1 (1965).
4. Ames, W.F.,Nonlinear Partial Differential Equations in Engineering, Vol.2,Academic Press (1972).
5. Seshadri, R. and Singh,M.C., "Group Invariance in Nonlinear Motion of Rods and Strings", J. of Acoustic Soc., Vol. 76, No.4, 1169-1174 (1984).

Chapter 8

SIMILARITY ANALYSIS OF WAVE PROPAGATION PROBLEMS

8.0 General

A wave is any recognizable feature of disturbance that is transferred from one part of the medium to another with a recognizable velocity of propagation. There are two main classes of wave motion that can arise in physical situations: (1) propagation of waves along the characteristics of the governing hyperbolic partial differential equation, and (2) non-characteristic wave propagation for which the wave does not move along the characteristics. In this chapter, we will discuss different aspects of similarity analysis and the role of group invariance, as they relate to wave-propagation problems.

In chapter 7, the classification of a second order quasilinear partial differential equation was reviewed. In the analysis of wave propagation problems, the equations can also be expressed in the first order form:

$$U_t + MU_x + N = 0 \tag{8.1}$$

where U and N are (n $\times$ 1) column vectors, and M is a (n $\times$ n) square matrix. The eigen values $\lambda^{\ell};\ell$=1,...,n of the equation, det(M-λI)=0, are real and distinct for totally hyperbolic systems. The characteristics of Eq.(8.1) are given by

$$C^{\ell} \ : \ (\frac{dx}{dt})_{\ell} = \lambda^{\ell} \tag{8.2}$$

Consider, as an example, the linear wave equation

$$u_{tt} - \gamma^2 u_{xx} = 0 \tag{8.3}$$

Based on the form as described in Eq.(7.1),

$$B^2 - 4AC = 4\gamma^2$$

which is greater than zero. Therefore, the equation is hyperbolic. The characteristics are given by the following:

$$\lambda^{(1)} = (\frac{dx}{dt})_1 = \frac{\sqrt{B^2 - 4AC}}{2A} = +\gamma \tag{8.4a}$$

$$\lambda^{(2)} = (\frac{dx}{dt})_2 = -\frac{\sqrt{B^2 - 4AC}}{2A} = -\gamma \tag{8.4b}$$

Alternatively, Eq.(8.3) can be rewritten in the first order form,Eq.(8.1), by letting u_x and u_t equal to v and w, respectively. Then,

$$w_t - \gamma^2 v_x = 0 \tag{8.5a}$$

$$v_t - w_x = 0 \tag{8.5b}$$

If the vector

$$U = \begin{pmatrix} v \\ w \end{pmatrix},$$

then Eq.(8.3) can be written as

$$\begin{pmatrix} v_t \\ w_t \end{pmatrix} + \begin{pmatrix} 0 & -1 \\ -\gamma^2 & 0 \end{pmatrix} \begin{pmatrix} v_x \\ w_x \end{pmatrix} = 0$$

The characteristics are obtained by solving

$$det(M - \lambda I) = \begin{pmatrix} -\lambda & -1 \\ -\gamma^2 & -\lambda \end{pmatrix} = 0 \tag{8.6}$$

Therefore, the λ's are given by

$$\lambda^{(1)} = +\gamma \quad ; \quad \lambda^{(2)} = -\gamma$$

which is the same result as Eq.(8.4).

For more information of equation classification, readers are referred to references [1] and [2].

8.1 Propagation Along Characteristics

A great number of wave propagation problems are governed by the quasilinear hyperbolic equation

$$\psi(x, t, u, u_x, u_t) u_{xx} - u_{tt} = 0 \tag{8.7}$$

The simplest model for wave propagation is expressible in the form

$$\rho_t + m(\rho)\rho_x = 0 \tag{8.8}$$

where $m(\rho)$ is a given function of the dependent variable. The characteristic is given by

$$\frac{dx}{dt} = m(\rho)$$

and different values of ρ would propagate with different speeds, $m(\rho)$. The dependence of m on ρ produces a typical nonlinear distortion of the wave as it propagates. When m' is less than zero, higher values of ρ propagate with slower speeds. However, when m' is greater than zero, higher values of ρ propagate faster than the lower ones. As the distortion progresses, the wave would ultimately break, resulting in the formation of a shock *,

* In the formation of Eq.(8.8), shock waves appear as discontinuities in ρ. The derivation of Eq.(8.8) involves approximations which are not strictly correct. For example, in gas dynamics, the corresponding approximation relates to the omission of viscous and heat conduction effects.

as shown in Fig.8.1. Generally speaking, shocks need not propagate along the characteristics. The propagation along the characteristics occur for $0 < t < t_B$, where t_B is the time for the breakdown of a wave. When $m(\rho)$ is a constant, then the wave is propagated at a constant speed without change of shape along the lines of characteristics.

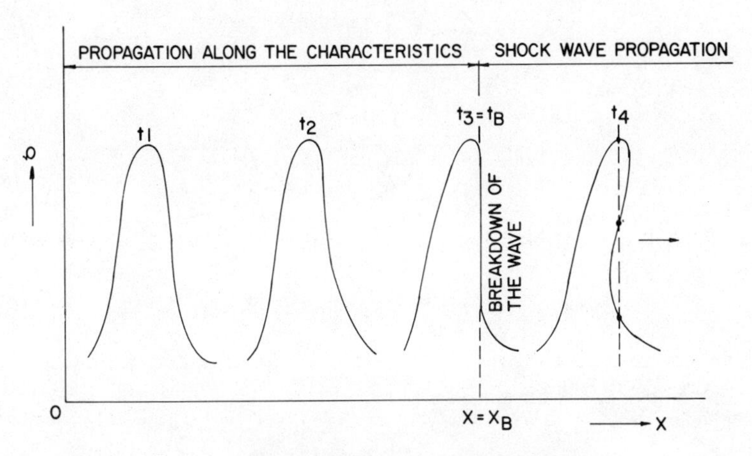

Fig.8.1 The Formation of Shock Waves

In this section, we will consider the analysis of the hyperbolic type equations for $0 < t < t_B$. It has been shown that the invariance of the governing equations for wave propagation problems under a group of transformations leads not only to the similarity transformations, but also to the equation's characteristics[3]. This aspect of group invariance is utilized in locating the moving boundary in terms of the similarity coordinate.

The characteristics for Eq.(8.7) can be written as

$$\lambda = \frac{dx}{dt} = \kappa(x, t, u, u_x, u_t) \tag{8.9}$$

We will now derive the similarity characteristic relationship which will essentially give the additional condition at the wave front that would be required to solve the similarity problem.

Case (1) : $\psi = \psi(x, t)$ *or* $\lambda = \kappa_1(x, t)$

Integrating along the characteristics, the position of the wave front can be expressed as

$$X_1(t) = \int \kappa_1(x, t)\, dt + X_0^{(1)} \tag{8.10}$$

where $X_0^{(1)}$ is a constant of integration. If $\varsigma_1 = \varsigma_1(x, t)$ is the similarity variable, then

$$\varsigma_1(x, t) = \varsigma_W^{(1)} \tag{8.11}$$

would locate the wavefront. $\varsigma_W^{(1)}$ can usually be determined from the equivalence of Eq.(8.10) and (8.11), which is known as the 'similarity characteristic' relationship. If

$$\psi(x,t) = A^* x^m t^n$$

in Eq.(8.7), then

$$\frac{dx}{dt} = \sqrt{A^*} x^{m/2} t^{n/2} \tag{8.12}$$

Since $X(0)=0$,

$$X(t) = (\sqrt{A^*}(\frac{2-m}{2+n}))^{\frac{2}{2-m}} t^{\frac{2+n}{2-m}} \tag{8.13}$$

The similarity variable for Eq.(8.7) with ψ $(x,t)=A^* x^m t^n$ can be written as

$$\varsigma_1(x,t) = \frac{x}{t^{(2+n)/(2-m)}} \tag{8.14}$$

The distance to the wavefront X(t) can be obtained from Eq.(8.14), by setting $\varsigma_1 = \varsigma_W^{(1)}$. Therefore,

$$X(t) = \varsigma_W^{(1)} t^{(2+n)/(2-m)} \tag{8.15}$$

From the equivalence of Eqs.(8.13) and (8.15) the similarity characteristic relationship can be obtained as

$$\varsigma_W^{(1)} = (\sqrt{A^*}\frac{2-m}{2+n})^{\frac{2}{2-m}}. \tag{8.16}$$

$\varsigma_W^{(1)}$ is the similarity coordinate at the wavefront.

Case (2) : $\psi = \psi(x,t,u,u_x,u_t)$

Therefore, the characteristics are given by

$$\lambda = \frac{dx}{dt} = \kappa_2(x,t,u,u_x,u_t) \tag{8.17}$$

Integration of Eq.(8.17) is not possible since there is a dependence on u, u_x, u_t, which are unknowns. However, by introducing the similarity transformation for Eq.(8.7) in conjunction with Eq.(8.17), i.e.,

$$u(x,t) = \beta(x,t)F(\varsigma_2),$$

Eq.(8.17) can be integrated as

$$X(t) = \int K_2[x,t,F(\varsigma_W^{(2)}),F'(\varsigma_W^{(2)})]dt + X_0^{(2)} \tag{8.18}$$

where $\varsigma_W^{(2)}$ is the similarity coordinate at the wavefront. $X_0^{(2)}$ can be evaluated by setting X(0)=0.

Setting the similarity variable equal to $\varsigma_W^{(2)}$,

$$\varsigma_2(x,t) = \varsigma_W^{(2)} \tag{8.19}$$

From the equivalence of Eqs.(8.18) and (8.19), $\varsigma_W^{(2)}$ can be determined.

Example 8.1: Nonlinear Wave Motion On a String

Consider the wave motion on a string with a gravitational force acting on a string in the negative axial direction. The equation can be expressed as [4]

$$u_{tt} - xu_{xx} - u_x = 0 \tag{8.20}$$

The similarity transformation can be written as

$$u(x,t) = t^a f(\varsigma) \quad where \quad \varsigma = \frac{x}{t^2} \tag{8.21}$$

The characteristics are given by

$$\frac{dx}{dt} = \pm\sqrt{x} \tag{8.22}$$

Choosing the positive characteristic and integrating Eq.(8.22), with $x(0)$=0:,

$$\sqrt{x} - \frac{t}{2} = 0 \;\; ; or \;\; X(t) = \frac{1}{4}t^2 \tag{8.23}$$

From Eq.(8.21), by setting $\varsigma = \varsigma_w$, we get

$$X(t) = \varsigma_w t^2 \tag{8.24}$$

Comparing Eqs.(8.23) and (8.24), the similarity coordinate at the wavefront is ς_w=1/4.

Example8.2 Rainfall Runoff in Sloping Areas

The physical problem considered here is one of a buildup of laminar or turbulent flow over a sloping area[5]. Consider an impermeable surface of length L, slope S (approximately equal to $sin(\theta)$) and of a unit width perpendicular to the plane as shown in Fig.8.2. Taking A as the origin, the two-dimensional flow will be examined. The continuity equation takes the form

$$\frac{\partial q}{\partial x} + \frac{\partial h}{\partial t} = v_0 \tag{8.25}$$

where q is the flow, h is the height of the water surface above the slope surface, and v_0 is the velocity of rainfall.

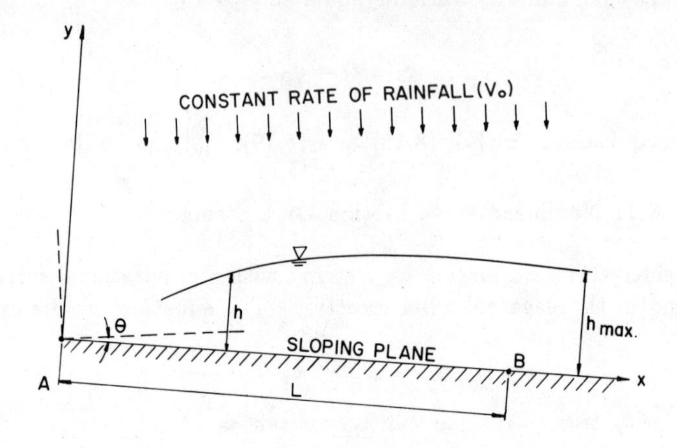

Fig.8.2 Rainfall On a Sloping Area

The flow and depth are related by the equation

$$q = \alpha h^m \tag{8.26}$$

where $\alpha = gS/3\nu$ and m=3 for laminar flow;g is the acceleration due to gravity and ν is the kinematic viscosity. For turbulent flow,

$$\alpha = C\sqrt{S} \;\; ; \;\; m = \frac{3}{2}$$

and C is a constant.

Eqs.(8.25) and (8.26) constitute a kinematic wave problem, and can be combined as

$$\alpha m h^{m-1} \frac{\partial h}{\partial x} + \frac{\partial h}{\partial t} = v_0 \tag{8.27}$$

The boundary condition is

$$h(0,t) = 0 \tag{8.28}$$

At the moving boundary,

$$h(X(t),t) = h_{max} = v_0 t$$

The similarity transformation is given by

$$h(x,t) = v_0 t f(\varsigma) \;\; where \;\; \varsigma = \frac{x}{\alpha m {v_0}^{m-1} t^m} \tag{8.29}$$

The ordinary differential equation obtained by substituting Eq.(8.29) into Eq.(8.27) is

$$(f^{m-1} - m\varsigma)\frac{df}{d\varsigma} + (f - 1) = 0 \tag{8.30}$$

with boundary condition $f(0)=0$.
The moving boundary is obtained by setting $\varsigma = \varsigma_W$, and can be written as

$$X(t) = \varsigma_w \alpha m v_0^{m-1} t^m \tag{8.31}$$

The characteristic of Eq.(8.27) is

$$\frac{dX}{dt} = m\alpha h^{m-1} = m\alpha[v_0 t f(\varsigma_w)]^{m-1} \tag{8.32}$$

Integrating Eq.(8.32) and comparing with Eq.(8.31),we get

$$\varsigma_w = \frac{1}{m} \tag{8.33}$$

Therefore, the propagation of the moving boundary is described by

$$X(t) = \alpha v_0^{m-1} t^m$$

The solution to Eq.(8.30) can be written as

$$f(\varsigma) = (m\varsigma)^{1/m} \tag{8.34}$$

Example8.3: Impact of a Longitudinal Rod With a Nonlinear Stress Strain Relationship[3]

In this example, one-dimensional deformation due to impact of a long thin rod with nonlinear stress strain relationship is considered. The governing equations for small deformations within the framework of the uniaxial theory of thin rods are as follows:

$$\frac{\partial \sigma}{\partial x} = -\rho \frac{\partial v}{\partial t} \tag{8.35a}$$

$$\frac{\partial e}{\partial t} = -\frac{\partial v}{\partial x} \tag{8.35b}$$

$$e = \left(\frac{\sigma}{\mu}\right)^q \tag{8.35c}$$

μ, ρ, q are material constants, x is the Lagrangian space coordinate, t is time, σ and e are normal compressive stress and nominal compressive strain respectively, v is the particle velocity and u is the particle displacement.

$$e = -\frac{\partial u}{\partial x} \quad and \quad v = \frac{\partial u}{\partial t}$$

The auxiliary conditions for the problem are:

$$v(0,t) = \frac{\partial u(0,t)}{\partial t} = v_c t^\delta \quad (v_c > 0;\ \delta \textit{ is a parameter})$$

Combining Eqs.(8.38), we have

$$\frac{\mu}{\rho q}[(-\frac{\partial u}{\partial x})^{(1-q)/q}\frac{\partial^2 u}{\partial x^2}] = \frac{\partial^2 u}{\partial t^2} \tag{8.36a}$$

The moving boundary is given by

$$u(D(t),t) = 0 \tag{8.36b}$$

where $x{=}D(t)$ locates the wavefront. The initial conditions are

$$u(x,0) = \frac{\partial u(x,0)}{\partial x} = 0 \tag{8.36c}$$

The similarity transformation can be written as

$$u(x,t) = v_c t^{\delta+1} F(\varsigma) \tag{8.37}$$

where

$$\varsigma = \frac{kx}{t^m} \quad ; \quad k = (\frac{\rho q}{\mu})^{\frac{q}{q+1}} (\frac{1}{v_c})^{\frac{1-q}{1+q}}$$

$$m = \frac{(1+\delta) + q(1-\delta)}{1+q}$$

The similarity representation can be found to be

$$[(-F')^{\frac{1-q}{q}} - m^2\varsigma^2]F'' - m(m-2\delta-1)\varsigma F'$$

$$-\delta(\delta+1)F = 0 \tag{8.38a}$$

with

$$F(0) = \frac{1}{1+\delta} \quad ; \quad F(\varsigma_w) = 0 \tag{8.38b, c}$$

where ς_w is the value of the similarity variable at the wavefront. Setting $\varsigma = \varsigma_w$, the wavefront description can be written as

$$D(t) = \varsigma_w \frac{t^m}{k} \tag{8.39a}$$

The characteristic is given by

$$\frac{dx}{dt} = \sqrt{\frac{\mu}{\rho q}}(-\frac{\partial u}{\partial x})^{\frac{1-q}{2q}} \tag{8.39b}$$

Introducing Eq.(8.37) into Eq.(8.39), integrating and then comparing with Eq.(8.39a),

$$\varsigma_w = \frac{[-F'(\varsigma_w)]^{(1-q)/2q}}{1+\delta(1-q)/(1+q)} \tag{8.40}$$

Eq.(8.40) should be satisfied at the wavefront. The reader may also note that the singularity of Eq.(8.38a) occurs when

$$(-F')^{\frac{1-q}{q}} - m^2\varsigma^2 = 0$$

which leads to the coordinate of the propagating discontinuity as expressed by Eq.(8.40).

Propagation of waves along the characteristics can occur in certain traveling-wave problems. The similarity representation would be a result of invariance of the governing equations and auxiliary conditions under a group of transformations:

$$G: \quad \bar{x} = x + \epsilon\alpha \ ; \ \bar{t} = t + \epsilon\beta \ ; \ \bar{u} = u$$

The traveling wave similarity solution can be written as

$$u = F(\varsigma) \ ; \ \varsigma = x - \lambda t \quad where \ \lambda = \frac{\alpha}{\beta} \tag{8.41}$$

is the speed of propagation of the wave.

Example8.4: The Linear Wave Equation

The linear wave equation can be written as

$$c^2\frac{\partial^2 u}{\partial x^2} = \frac{\partial^2 u}{\partial t^2} \tag{8.42}$$

c is the velocity of propagation of the wave.

Eq.(8.42) is invariant under Eq.(8.41), and the similarity representation is

$$(c^2 - \lambda^2)F'' = 0 \tag{8.43}$$

where $\lambda = \pm c$ are the speeds of propagation of traveling waves.

8.2 Non-Characteristic Propagation: Shock Waves

In section 8.1, the breakdown of a smooth wave propagating along the characteristics into shocks had been discussed. Shock propagation is generally a non-characteristic type of propagation. The breakdown of waves into shocks can result from certain forms of constitutive relationships, or it can result from high impact or high energy release in a medium. Instead of using the similarity characteristic relationship, the so-called "jump conditions" have to be satisfied at the shock front.

For the case of uniaxial rods for which the governing equations have been described in section 8.1, Eqs.(8.35a) and (8.35b), the jump conditions are obtained from the continuity of displacement and momentum as follows:

$$[v] = -c[e] \quad ; \quad c\rho[v] = -[\sigma] \tag{8.44a, b}$$

where c is the as yet unknown velocity of the shock front, and is equal to:

$$c = \sqrt{\frac{1}{\rho}\frac{[\sigma]}{[e]}}$$

These relations together with the constitutive equation are sufficient for the study of shock propagation in thin rods. When the impact velocity is high, then the variation of internal energy would be a factor. Eqs.(8.44a) and (8.44b) are called the Hugonoit conditions. The symbol [ ] denotes the difference in the variable across the shock front.

For a compressible fluid, the velocity, pressure,density and temperature of the fluid are discontinuous across the shock. For an ideal gas, the relationships for the variables across the shock are:

$$Continuity \ : \quad \rho_1 u_1 = \rho_2 u_2 \tag{8.45a}$$

$$Momentum \ : \ \rho_1 u_1(u_1 - u_2) = p_2 - p_1 \tag{8.45b}$$

$$Energy \ : \ (\frac{\gamma}{\gamma-1})\frac{p_1}{\rho_1} + \frac{u_1^2}{2} = (\frac{\gamma}{\gamma-1})\frac{p_2}{\rho_2} + \frac{u_2^2}{2} \tag{8.45c}$$

where u is the velocity, p is the pressure, and ρ is density. Subscripts *1* refers to the values of the variables in front of the shock, and *2* behind the shock(Fig.8.3).

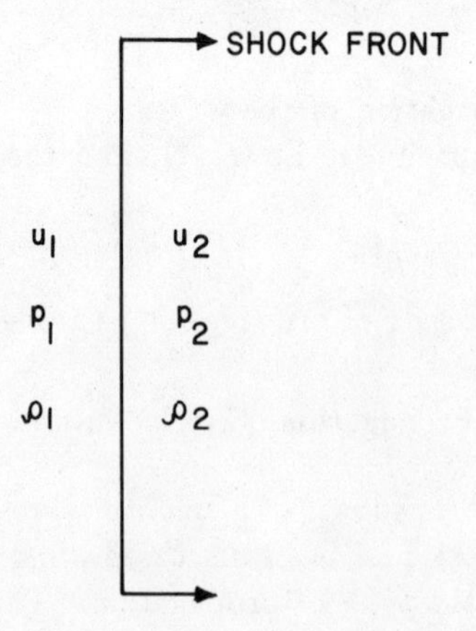

Fig. 8.3 Shock in a Compressible Media

Example 8.5 Formation of a Blast Wave by an Intense Explosion[2.6]

When a finite amount of energy is suddenly released in an infinitely concentrated form,the resulting motion in the air medium can be determined by similarity analysis. If the disturbance is strong enough, the initial pressure and sound speed of ambient air are negligible compared to the pressures and velocities produced in the disturbed flow.

The equations describing the blast wave phenomena are:

$$\rho_t + u\rho_r + \rho(u_r + \beta\frac{u}{r}) = 0 \tag{8.46a}$$

$$u_t + uu_r + \frac{1}{\rho}p_r = 0 \tag{8.46b}$$

$$p_t + up_r - a^2(\rho_t + u\rho_r) = 0 \tag{8.46c}$$

ρ is the density,p is the pressure, u is the radial velocity,r is the radius, t is the time, and a is the speed of sound defined as

$$a = \sqrt{\frac{\gamma p}{\rho}}$$

When β equals to *1* the motion is cylindrical, and when β equals to *2* the motion is spherical. The total energy, E, which is conserved, can be written for $\beta = 2$ as:

$$E = \int_0^{R(t)} (\frac{p}{\gamma - 1} + \frac{1}{2}\rho u^2) 4\pi r^2 \, dr \tag{8.47}$$

At the shock front, instead of the similarity characteristic relationship the jump conditions are utilized as follows:

$$u(R,t) = \frac{2U}{\gamma + 1} \tag{8.48a}$$

$$\rho(R,t) = (\frac{\gamma + 1}{\gamma - 1})\rho_0 \tag{8.48b}$$

$$p(R,t) = \frac{2\rho_0 U^2}{\gamma + 1} \tag{8.48c}$$

where U is the shock velocity,ρ_0 is the ambient gas density and γ is the polytropic exponent.

The similarity transformation can be obtained as [2.6]

$$u(r,t) = \frac{\alpha r}{t} V(\varsigma) \tag{8.49a}$$

$$\rho(r,t) = \rho_0 Q(\varsigma) \tag{8.49b}$$

$$p(r,t) = \alpha^2 \rho_0 \frac{r^2}{t^2} P(\varsigma) \tag{8.49c}$$

where

$$\varsigma = \frac{r}{(Ct)^\alpha}$$

We can determine C such that $R(\mathrm{t}){=}(Ct)^\alpha$ is the moving boundary. Letting

$$A = \sqrt{\frac{\gamma P}{Q}},$$

the similarity representation can now be written as:

$$[(V-1)^2 - A^2]\varsigma V' = [(\beta+1)V - \frac{2(1-\alpha)}{\alpha\gamma}]A^2$$

$$-V(V-1)(V-\frac{1}{\alpha}),$$

$$[(V-1)^2 - A^2]\frac{\varsigma A'}{A} = [1 - \frac{(1-\alpha)}{\gamma\alpha}\frac{1}{V-1}]A^2$$

$$+(\frac{\gamma-1}{2})V(V-\frac{1}{\alpha}) - (\frac{\gamma-1}{2})(\beta+1)V(V-1) - (V-1)(V-\frac{1}{\alpha}),$$

$$[(V-1)^2 - A^2]\frac{\varsigma Q'}{Q} = 2[(\beta+1)V - (\frac{1-\alpha}{\gamma\alpha})]\frac{A^2}{(V-1)}$$

$$-V(V-\frac{1}{\alpha}) - (\beta+1)V(V-1). \tag{8.50a, b, c}$$

The velocity of the shock is

$$U = \frac{dR}{dt} = \frac{\alpha R}{t} = \frac{2R}{5t}$$

The transformed conditions at the shock front are

$$V(1) = \frac{2}{\gamma+1} \;;\; Q(1) = \frac{\gamma+1}{\gamma-1} \;;\; P(1) = \frac{2}{\gamma+1} \tag{8.51}$$

The constant C can be obtained from the energy integral, Eq.(8.47). For $\gamma = 1.4$, $\beta{=}2$, the description of the motion for the shock front is

$$S(t) = 1.025(\frac{E}{\rho_0})^{1/5} t^{2/5} \tag{8.52}$$

We will now examine the characteristics for Eq.(8.46) and relate it to the path of the shock front.

The characteristics in the $r - t$ plane is given by

$$\frac{dr}{dt} = u \pm a \tag{8.53}$$

Since u and a are given by

$$u = \frac{\alpha r}{t} V(\varsigma) \quad and \quad a = \frac{\alpha r}{t} A(\varsigma),$$

we get

$$\frac{dr}{dt} = \frac{\alpha r}{t} [V(\varsigma_0) \pm A(\varsigma_0)] \tag{8.54}$$

where ς_0 is the similarity variable at the characteristics. Eq.(8.54) can be integrated as

$$R(t) = R_0 t^{\alpha [V(\varsigma_0) \pm A(\varsigma_0)]} \tag{8.55}$$

where R_0 is a constant.

In terms of the similarity variable, $\varsigma = \varsigma_0$ would describe characteristic propagation such that

$$R(t) = \varsigma_0 C^{\alpha} t^{\alpha} \tag{8.56}$$

Comparing Eqs.(8.55) and (8.56),

$$\varsigma_0 = \frac{R_0}{C^{\alpha}} \quad and \quad V(\alpha_0) \pm A(\varsigma_0) = 1$$

In general, α is not equal to 2/5, and ς_0 is different from one. Therefore, the shock path and characteristics do not coincide in general (Fig.8.4).

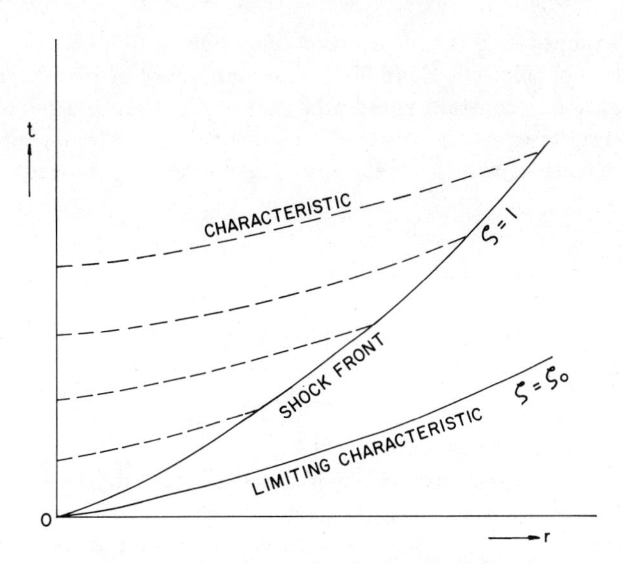

Fig.8.4 Plot for the Explosion Problem

The limiting characteristics appears in the region ahead of the shock. It represents the edge of the fold in the (r, t) plane in the multivalued solution which is replaced by the shock.

8.3 Non-characteristics Propagation: Uniform Propagation Regime

In the previous section, similarity analysis of problems involving propagation of shocks was discussed. It was seen that the moving boundary does not, in general, propagate along the equation's characteristics. Instead of the "similarity characteristic"relationship, the "jump conditions" are required to locate the moving boundary.

The propagation of "traveling waves" is another possible situation, where the moving boundary need not propagate along the characteristics, especially if the equations are nonlinear. As discussed in section 5.3, traveling wave solutions can be obtained by invoking invariance under a group of translations

$$G: \quad \bar{x} = x + \epsilon\alpha \;\; ; \;\; \bar{t} = t + \epsilon\beta \;\; ; \;\; \bar{u} = u$$

Invariant solutions can be obtained by solving the subsystem corresponding to G,i.e.,

$$\frac{dx}{\alpha} = \frac{dt}{\beta} = \frac{du}{0},$$

and can be written as:

$$u = F(\varsigma) \;\; ; \;\; \varsigma = x - \lambda t \tag{8.57}$$

where the assumed velocity of propagation λ equals to α/β.

The class of solutions,Eq.(8.57), represent waves of permanent profile that propagate at a constant speed and unchanging shape. The reduced ordinary differential equation obtained on substituting Eq.(8.57) into a given hyperbolic equation, will then represent possible modes of steady propagation.

Example8.6: Klein-Gordon Equation

A mechanical transmission line treated by Scott[7] is modeled by the following equation:

$$\phi_{xx} - \phi_{tt} = G(\phi) \tag{8.58a}$$

Eq.(8.58) is known as the Klein-Gordon equation.

Scott describes his construction of a mechanical model with rigid pendula attached at close intervals along a stretched wire. Torsional waves propagating down the wire obey the wave equation, and the pendula supply a restoring force proportional to $\sin\phi$, where ϕ is the angular displacement.

Therefore, $G(\phi)=sin\,\phi$, and Eq.(8.58) which is called the Sine-Gordon equation can be written as:

$$\phi_{xx} - \phi_{tt} = sin\,\phi \tag{8.58b}$$

Since Eq.(8.58b) is invariant under a group of translations

$$G: \bar{x} = x + \epsilon\alpha \;\; ; \;\; \bar{t} = t + \beta \;\; ; \;\; \bar{\varphi} = \varphi$$

Eq.(8.57) is the required traveling wave similarity solution. Therefore,

$$\phi = \phi(\varsigma) \; and \; \varsigma = x - \lambda t$$

Eq.(8.58b) then becomes

$$(1-\lambda^2)\phi'' = G(\phi) \tag{8.59}$$

Two distinct pulse solutions are obtained [8],namely,

$$\phi(\varsigma) = 4Tan^{-1}exp[\pm\frac{\varsigma}{\sqrt{1-\lambda^2}}] \;\; for \;\; \lambda < 1 \tag{8.60a}$$

and

$$\phi(\varsigma) = 4Tan^{-1}exp[\pm\frac{\varsigma}{\sqrt{\lambda^2-1}}] + \pi \;\; for \;\; \lambda > 1 \tag{8.60b}$$

In the first case, the constant of integration is set equal to +1, and in the second case, -1. If the integration constant (c_0) is not equal to ± 1,ϕ can be written as an implicit function of the similarity variable,ς, given by

$$\sqrt{1-\lambda^2}\int_0^{\phi}\frac{d\phi}{\sqrt{2(c_0 - cos\,\phi)}} = \varsigma \tag{8.61}$$

For the case $c_0 > 1$,$\lambda < 1$,ϕ is a monotonically increasing function of ς:

$$\phi(\varsigma) = Cos^{-1}\left[2cd^2(\frac{\varsigma}{\gamma(1-\lambda^2)^{1/2}}) - 1\right] \tag{8.62}$$

where cd x=(cn x)/(dn x) is an elliptic function of modulus $\gamma = 2(c_0+1)$.

For the second case of interest for which $-1 < c_0 < 1$ and $\lambda > 1$, $\phi(\varsigma)$ is a periodic function of ς.

$$\phi(\varsigma) = 2\;sin^{-1}\left[\gamma\;sn(\frac{\varsigma}{\sqrt{\lambda^2-1}})\right] \tag{8.63}$$

where *sn* is an elliptic function of modulus $2\gamma^2 = 1 - c_0$.

Example8.7 Korteweg-de Vries Equation

In connection with the study of water waves of permanent profile, the Korteweg-de Vries equation arises in the form [2]:

$$\eta_t + c_0\left(1 + \frac{3}{2}\frac{\eta}{h_0}\right)\eta_x + \gamma\eta_{xxx} = 0 \tag{8.64}$$

where η is the water surface position, h_0 locates the bottom,and c_0 and γ are defined by

$$c_0 = \sqrt{gh_0} \quad and \quad \gamma = \frac{1}{6}c_0 h_0^2$$

Again, Eq.(8.64) is invariant under a group of translations:

$$G: \quad \bar{x} = x + \epsilon\alpha \;\; ; \;\; \bar{t} = t + \epsilon\beta \;\; ; \;\; \bar{\eta} = \eta$$

so that a traveling-wave similarity solution can be written as:

$$\eta = h_0\varphi(\varsigma) \;\; ; \;\; \varsigma = x - Ut$$

Therefore, the similarity representation is given by

$$\frac{1}{6}h_0^2\varphi''' + \frac{3}{2}\varphi\varphi' - \left(\frac{U}{c_0} - 1\right)\varphi' = 0 \tag{8.65}$$

This can be integrated to

$$\frac{1}{6}h_0^2\varphi'' + \frac{3}{4}\varphi^2 - \left(\frac{U}{c_0} - 1\right)\varphi + A = 0$$

After multiplying by φ', further integration gives

$$\frac{1}{3}h_0^2(\varphi')^2 + \varphi^3 - 2\left(\frac{U}{c_0} - 1\right)\varphi^2 + 4G\varphi + B = 0 \tag{8.66}$$

where A and B are constants of integration.

In the special situation when φ and its derivatives become zero as

$$\varsigma \rightarrow \infty \; : \; A = B = 0$$

Therefore, Eq.(8.66) may be rewritten as

$$\frac{1}{3}h_0^2\left(\frac{d\varphi}{d\varsigma}\right)^2 = \varphi^2(\alpha - \varphi) \quad where \quad \frac{U}{c_0} = 1 + \frac{\alpha}{2} \tag{8.67}$$

It is clear, at least qualitatively, that φ increases from $\varphi = 0$ at $\varsigma = \infty$ and rises to a maximum at $\varphi = \alpha$. It then drops to $\varphi = 0$ at $\varsigma = -\infty$.

This is a "solitary wave", the velocity of which depends on the amplitude according to the relationship

$$U = c_0\left(1 + \frac{1}{2}\frac{\eta_0}{h_0}\right) \quad where \quad \eta_0 = h_0\alpha$$

The solution of Eq.(8.67) is (see Fig.8.5):

$$\varphi(\varsigma) = \alpha \ sech^2\left[\sqrt{\frac{3\alpha}{4h_0^2}}\varsigma\right] \tag{8.68a}$$

or in terms of the original variables,

$$\eta(x,t) = \eta_0 Sech^2\left[\sqrt{\frac{3\eta_0}{4h_0^3}}(x - Ut)\right] \tag{8.68b}$$

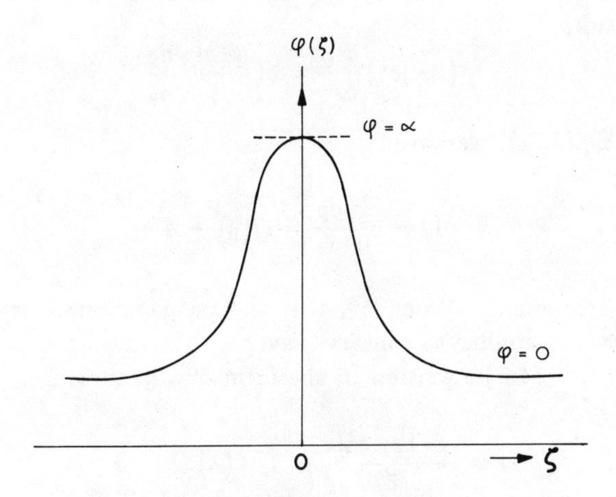

Fig. 8.5 Solitary Wave

Example8.8: Elasto-plastic Wave Propagation in a Rod

For a constitutive relationship described by

$$\frac{\partial e}{\partial t} = \frac{1}{E}\frac{\partial \sigma}{\partial t} + \gamma f(\sigma),$$

the equation for one-dimensional wave propagation in a rod is given by [10]:

$$\beta\left(\frac{\partial^2\sigma}{\partial\xi^2} - \frac{\partial^2\sigma}{\partial\tau^2}\right) = \frac{df(\sigma)}{d\sigma}\frac{\partial\sigma}{\partial\tau} \tag{8.69}$$

where x is the coordinate along the rod, t is the time, e is strain, σ is the overstress, E is the modulus of elasticity, γ is a constant, σ_0 is static yield stress, and

$$\xi = \alpha x \ ; \ \tau = \alpha c t \ ; \ c = \sqrt{\frac{E}{\rho}} \ ; \ \alpha = \beta\gamma\sqrt{\frac{\rho E}{\sigma_0}}$$

Eq.(8.69) is invariant under a group of translations,

$$G : \ \bar{\xi} = \xi + \epsilon\alpha \ ; \ \bar{\tau} = \tau + \epsilon\beta \ ; \ \bar{\sigma} = \sigma,$$

so that a traveling wave solution can be written as

$$\sigma = g(s) \ \ and \ \ s = \bar{c}\tau - \xi \tag{8.70}$$

where $\bar{c}$ is the speed of propagation, equal to α/β.

Substituting Eq.(8.70) into Eq.(8.69), the following similarity representation results:

$$\beta(1-\bar{c}^2)\frac{d^2g}{ds^2} = \bar{c}\left(\frac{df(g)}{dg}\right)\frac{dg}{ds} \tag{8.71}$$

Integrating Eq.(8.71), we have

$$\beta(1-\bar{c}^2)\frac{dg}{ds} = \bar{c}f(g) + A$$

where A is a constant. When $\bar{c}^2 < 1$, the propagation of waves is non-characteristic and similar to solitary waves.

The solution can be written in the form of a quadrature:

$$s = \left[\frac{\beta(1-\bar{c}^2)}{\bar{c}}\right]\int^{g} f^{-1}(\lambda)\,d\lambda + B \tag{8.72}$$

The constant A is set equal to zero, and B is another constant. Once the form of $f(g)$ is known, Eq.(8.72) can be evaluated.

8.4 From Translation to Dimensional Group Invariance

In section 8.3, the traveling wave similarity solutions were discussed. The equations are invariant under a group of translations, and the similarity solutions are of the form:

$$u = u(\varsigma) \ \ ; \ \ \varsigma = x - \lambda t - c \tag{8.73}$$

where λ is the propagation speed, and c is a constant.

We now make a change of variables as follows:

$$x = ln(\xi) \quad ; \quad t = ln(\tau) \quad ; \quad c = ln(A) \tag{8.74}$$

Eq.(8.73) can then be written as

$$u = U(\frac{\xi}{A\tau^{\lambda}}) \tag{8.75}$$

which is the type of similarity variable obtained by dimensional groups.

*Example*8.9: Korteweg-de Vries Equation

We will rewrite the Korteweg-de Vries Equation in a different form:

$$u_{xxx} + uu_x + u_t = 0 \tag{8.76}$$

By making the change of variables as expressed by Eq.(8.74), the Korteweg-de Vries equation can be transformed to:

$$\xi^3 \frac{\partial^3 u}{\partial \xi^3} + 3\xi^2 \frac{\partial^2 u}{\partial \xi^2} + \xi \frac{\partial u}{\partial \xi} + \xi u \frac{\partial u}{\partial \xi} + \tau \frac{\partial u}{\partial \tau} = 0 \tag{8.77}$$

Since Eq.(8.77) is a similarity representation of Eq.(8.76), it can be transformed to an ordinary differential equation in terms of the variable θ where

$$\theta = \frac{\xi}{A\tau^{\lambda}}$$

When traveling-wave solutions of the form Eq.(8.73), have an unknown velocity of propagation,λ,the problem basically corresponds to "self-similar solutions of the second kind"[11]. λ is determined from the simultaneous consideration of the conservation laws and the internal structure of the transition regime. More discussion on "self-similar solutions of the first and second kind" is presented in section 11.5 of this book.

8.5 Summary

In this chapter, similarity analysis of wave propagation problems governed by hyperbolic equations was discussed. Two classes of wave motion were considered: (1) propagation along the equations' characteristics, and (2) non-characteristic propagation. The analysis of the first class of wave motion involves the determination of the similarity characteristic relationship. This relationship essentially represents a singularity in the similarity representation. The second class of wave motion arises as a result of shock wave propagation and in traveling wave problems. Instead of satisfying the similarity characteristic relationship at the wave front, the so-called "jump conditions" are required for the shock wave propagation problem. The traveling wave solution is obtained by invoking invariance under a group of translations.

REFERENCES

1. Jeffrey, A. and Taniuti, T.,"Nonlinear Wave Propagation With Applications to Physics and Magneto-Hydrodynamics", Academic Press, New York (1964).
2. Whitham, G.B., Linear and Nonl. Waves, Wiley, (1974).
3. Seshadri, R. and Singh, M.C., "Similarity Analysis of Wave Propa. Probs. in Nonlinear Rods, Archives of Mechanics, No.6 (1980).
4. Seshadri, R. and Singh, M.C., "Group Invariance in Nonlinear Motion of Strings and Rods", Rept. 260, Dept. of Mech. Eng., Univ. of Calgary, Canada (1983).
5. Seshadri, R. and Jagannathan, J.R.,"Similarity Solution for Rainfall-Runoff Problem in Sloping Areas", Ind. Math., Vol.32, Part 2 (1982).
6. Taylor,G.I.,"The Formation of a Blast Wave by a Very Intense Explosion", Theoretical Discussion, Proc. Roy. Soc., A, 201 (1950).
7. Scott, A.C., Active and Nonlinear Wave Propagation in Electronics, Wiley-Interscience, New York (1970).
8. Ames, W.F., Nonlinear Partial Differential Equations in Engineering, Vol.2, Academic Press (1972).
9. Abramowitz, M. and Stegun, I.A.,"Handbook of Mathematical Functions", Dover Publications Inc. (1965).
10. Burniston, E.E. and Chang, T.S.,"Nonlinear Waves in Rate-Sensitive, Elastoplastic Material", Int'l J. Eng. Sci., Vol.10 (1972).
11. Barenblatt, G.I. and Zel'dovich, Ya. B.,"Self-Similar Solutions Are Intermediate Symptotics", Ann. Rev. of Fluid Mechanics (1972).
12. Sedov, L.I., Similarity and Dimensional Method in Mechanics, Academic Press (1959).

Chapter 9

TRANSFORMATION OF A BOUNDARY VALUE PROBLEM TO AN INITIAL VALUE PROBLEM

9.0 Introduction

The method for transforming nonlinear boundary value problems to initial value problems was first introduced by Toepfer[1] in 1912 in his attempt to solve Blasius' equation in boundary layer theory by a series expansion method. About half a century later Klamkin[2], based on the same reasoning, extended the method to a wider class of problems. Major extensions were made possible only when the transformation process was interpreted and re-examined by Na[3] in terms of the continuous groups of transformations.

In this chapter, two methods for transforming a boundary value problem into an initial value problem are discussed. The first method uses inspectional groups, and as such will be referred to as the "inspectional group method". The second method,which is deductive in nature, is based on the use of infinitesimal groups and will therefore be called the "infinitesimal group method". Both the methods start out by defining a group of transformations. The "particular transformation" within this group of transformations which can convert the boundary value problem into an initial value problem is identified. In the inspectional group method, the particular transformation within this group of transformations which can convert the boundary value problem into initial value problem is similarly identified. This is done by stipulating that:

(1) the given differential equation be independent of the parameter of transformation, and
(2) the parameter of transformation is identified as the "missing" boundary condition.

In the infinitesimal group method, invocation of invariance of the differential equation leads to a particular form of the characteristic function, W. This will result in a subsystem of equations which upon integrating from the variables in the boundary value problem to the variables in the initial value problem, gives the required transformation.

We will now examine the inspectional group method and the infinitesimal group method through applications to some engineering boundary value problems.

9.1 Blasius Equation in Boundary Layer Flow

As a first example, let us consider the Blasius equation from the boundary layer theory[4]. The objective is to transform the boundary value problem

$$\frac{d^3f}{d\eta^3} + \frac{1}{2}f\frac{d^2f}{d\eta^2} = 0 \tag{9.1}$$

$$\eta = 0 \ : \ f(0) = 0 \ , \quad \frac{df(0)}{d\eta} = 0 \tag{9.2a, b}$$

$$\eta = \infty \ : \ \frac{df(\infty)}{d\eta} = 1 \tag{9.2c}$$

into an initial value problem

$$\frac{d^3g}{d\xi^3} + \frac{1}{2}g\frac{d^2g}{d\xi^2} = 0 \tag{9.3}$$

$$\xi = 0 \ : \ g(0) = 0 \ ; \quad \frac{dg(0)}{d\xi} = 0 \ ; \quad \frac{d^2g(0)}{d\xi^2} = 1 \tag{9.4a, b, c}$$

We will now derive the transformation using both methods.

(a) Inspectional Group Method:

Defining a linear group of transformations, G:

$$G \ : \quad \eta = A^{\alpha_1}\xi \ \ and \ \ f = A^{\alpha_2}g \tag{9.5}$$

Transforming Eq.(9.1) by using group, G, we have

$$\frac{A^{\alpha_2}}{A^{3\alpha_1}}\frac{d^3g}{d\xi^3} + \frac{A^{2\alpha_2}}{A^{2\alpha_1}}\frac{1}{2}g\frac{d^2g}{d\xi^2} = 0 \tag{9.6}$$

Eqs.(9.1) and (9.3) would be equivalent if

$$\alpha_2 - 3\alpha_1 = 2\alpha_2 - 2\alpha_1 \tag{9.7a}$$

or

$$\alpha_2 = -\alpha_1 \tag{9.7b}$$

The first two boundary conditions, Eqs.(9.2a,b), would transform under Eq.(9.5) to the conditions (9.4a,b). However, the boundary condition, Eq.(9.2c) becomes

$$\frac{df(\infty)}{d\eta} = \frac{A^{\alpha_2}}{A^{\alpha_1}}\frac{dg(\infty)}{d\xi} = 1 \tag{9.8a}$$

Using the relationship as obtained in Eq.(9.7b),

$$A = \left(\frac{dg(\infty)}{d\xi}\right)^{-1/(2\alpha_2)} \tag{9.8b}$$

We now have to determine A such that the "stipulated" boundary condition, Eq.(9.4c) is obtained. Therefore,

$$\frac{d^2g(0)}{d\xi^2} = \frac{A^{2\alpha_1}}{A^{\alpha_2}} \frac{d^2f(0)}{d\eta^2} = 1 \tag{9.9a}$$

so that

$$A = \left(\frac{d^2f(0)}{d\eta^2}\right)^{1/(3\alpha_2)} \tag{9.9b}$$

We now set $f''(0)$ equal to the value of the parameter, A, so that Eq.(9.9b) leads to $\alpha_2 = 1/3$. Therefore, using Eq.(9.7b) $\alpha_1 = -1/3$.

Eliminating the parameter A such that Eq.(9.8b) and (9.9b) are equivalent, we have

$$A = \frac{d^2f(0)}{d\eta^2} = \left(\frac{dg(\infty)}{d\xi}\right)^{-3/2} \tag{9.10}$$

The numerical integration is now straightforward:

(i) Solve Eq.(9.3) subject to boundary conditions, Eq.(9.4), as an initial value problem by forward integration. Obtain the value of g'(∞).

(ii) Compute the value of A from the relationship

$$A = \left(\frac{dg(\infty)}{d\xi}\right)^{-3/2}$$

(iii) Using the transformation as defined by Eq.(9.5), determine the variation of f(η) vs. η. Therefore, the problem is now transformed back into its original description.

For the Blasius problem, the variation of $dg/d\xi$ vs. ξ is plotted in Fig.9.1. It can be seen from the figure that $dg(\infty)/d\xi$=2.0852. Therefore, the value of A from Eq.(9.10) is equal to 0.3320. The transformation can now be written as

$$G\ :\ \eta = A^{-1/3}\xi\ ;\ f = A^{1/3}g \quad where\ A = 0.3320. \tag{9.5b}$$

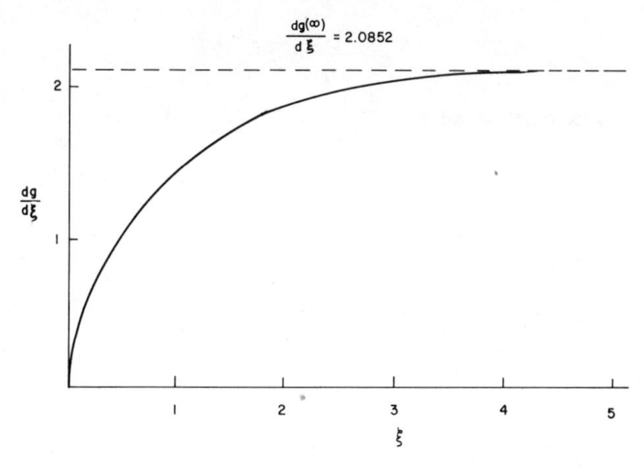

Fig.9.1 Solution of Eq.(9.3)

(b) Infinitesimal Group Method

The basic difference between this method and the inspectional method is that no particular group of transformation is defined at the outset. The required boundary to initial value transformation is derived systematically. To begin, let us define an infinitesimal group of transformations:

$$\begin{aligned}
\bar{\eta} &= \eta + \epsilon\phi(\eta, f) + O(\epsilon^2) \\
\bar{f} &= f + \epsilon\theta(\eta, f) + O(\epsilon^2) \\
\bar{p} &= p + \epsilon\varsigma(\eta, f, p) + O(\epsilon^2) \\
\bar{q} &= q + \epsilon\delta(\eta, f, p) + O(\epsilon^2) \\
\bar{r} &= r + \epsilon\rho(\eta, f, p) + O(\epsilon^2)
\end{aligned} \tag{9.11}$$

where

$$p = \frac{df}{d\eta} \ ; \ q = \frac{d^2 f}{d\eta^2} \ ; \ r = \frac{d^3 f}{d\eta^3}$$

In terms of the characteristic function, W,

$$\phi = \frac{\partial W}{\partial p} \ ; \ \theta = p\frac{\partial W}{\partial p} - W \ ; \ \varsigma = -X(W)$$

$$\delta = -\left[X^2 W + 2qX\frac{\partial W}{\partial p} + q\frac{\partial W}{\partial f}\right] \tag{9.12}$$

$$\rho = -\left[X^3 W + 3qX^2\frac{\partial W}{\partial p} + 3qX\frac{\partial W}{\partial f} + 3q^2\frac{\partial^2 W}{\partial f \partial p}\right]$$

$$-r\left[3X\frac{\partial W}{\partial p} + \frac{\partial W}{\partial f}\right]$$

with

$$X(\ ) = \frac{\partial(\ )}{\partial\eta} + p\frac{\partial(\ )}{\partial f}$$

Eq.(9.1) can be written as:

$$r + \frac{1}{2}fq = 0 \tag{9.13}$$

Invariance of Eq.(9.1) under transformation (9.11) can be written as

$$\rho + \frac{1}{2}f\delta + \frac{1}{2}q\theta = 0 \tag{9.14}$$

Using Eqs.(9.12), we get:

$$X^3 W + 3qX^2\frac{\partial W}{\partial p} + 3qX\frac{\partial W}{\partial f} + 3q^2\frac{\partial^2 W}{\partial f \partial p} - \frac{3}{2}fqX\frac{\partial W}{\partial p}$$

$$-\frac{1}{2}fq\frac{\partial W}{\partial f}+\frac{1}{2}f(X^2W+2qX\frac{\partial W}{\partial p}+q\frac{\partial W}{\partial f})$$

$$+\frac{q}{2}(W-p\frac{\partial W}{\partial p})=0 \tag{9.15}$$

Simplifying and re-writing Eq.(9.15), we have

$$X^3W+(3qX-\frac{1}{2}fq)X\frac{\partial W}{\partial p}+3qX\frac{\partial W}{\partial f}+3q^2\frac{\partial^2 W}{\partial f\partial p}$$

$$+\frac{1}{2}fX^2W+\frac{1}{2}qW-\frac{1}{2}pq\frac{\partial W}{\partial p}=0 \tag{9.16}$$

Since W is linear in p,

$$W=W_1(\eta,f)p+W_2(\eta,f) \tag{9.17}$$

Substituting Eq.(9.17) into Eq.(9.16), we get

$$\frac{\partial^3 W_2}{\partial\eta^3}+\frac{f}{2}\frac{\partial^2 W_2}{\partial\eta^2}+p(\frac{\partial^2 W_1}{\partial\eta^3}+3\frac{\partial^3 W_2}{\partial\eta^2\partial f}+\frac{f}{2}\frac{\partial^2 W_1}{\partial\eta^2}+f\frac{\partial^2 W_2}{\partial\eta\partial f})$$

$$+p^2(3\frac{\partial^3 W_1}{\partial f\partial\eta^2}+3\frac{\partial^3 W_2}{\partial\eta\partial f^2}+f\frac{\partial^2 W_1}{\partial\eta\partial f}+\frac{f}{2}\frac{\partial^2 W_2}{\partial f^2})$$

$$+p^3(3\frac{\partial^3 W_1}{\partial\eta\partial f^2}+\frac{\partial^3 W_2}{\partial f^3}+\frac{f}{2}\frac{\partial^2 W_1}{\partial f^2})+p^4\frac{\partial^3 W_1}{\partial f^3}$$

$$+q[3\frac{\partial^2 W_1}{\partial\eta^2}-\frac{f}{2}\frac{\partial W_1}{\partial\eta}+3\frac{\partial^2 W_2}{\partial\eta\partial f}+\frac{W_2}{2}]$$

$$+pq[-\frac{f}{2}\frac{\partial W_1}{\partial f}+9\frac{\partial^2 W_1}{\partial\eta\partial f}+3\frac{\partial^2 W_2}{\partial f^2}]$$

$$+6p^2q\frac{\partial^2 W_1}{\partial f^2}+3q^2\frac{\partial W_1}{\partial f}=0 \tag{9.18}$$

Equating the coefficients equal to zero, we can conclude based on the coefficient of q^2 that

$$W_1=W_1(\eta)$$

The coefficient of pq would lead to the conclusion that W_2 is linear in f.

The remaining coefficients can be written as:

$$\frac{\partial^3 W_2}{\partial\eta^3}+\frac{1}{2}f\frac{\partial^2 W_2}{\partial\eta^2}=0$$

$$\frac{\partial^3 W_1}{\partial\eta^3}+3\frac{\partial^3 W_2}{\partial f\partial\eta^2}+\frac{1}{2}f\frac{\partial^2 W_1}{\partial\eta^2}+f\frac{\partial^2 W_2}{\partial\eta\partial f}=0 \tag{9.19}$$

$$3\frac{\partial^2 W_1}{\partial \eta^2} - \frac{f}{2}\frac{\partial W_1}{\partial \eta} + 3\frac{\partial^2 W_2}{\partial \eta \partial f} + \frac{W_2}{2} = 0$$

The solution of Eqs.(9.19) would lead to:

$$W(\eta, f, p) = (c_5\eta + c_6)p + c_5 f - 6c_5 \tag{9.20}$$

The infinitesimals ς and δ are:

$$\varsigma = -2c_5 p \;\; ; \;\; \delta = -3c_5 q \tag{9.21}$$

The difference between the boundary value problem, Eqs.(9.1) and (9.2), and the initial value problem, Eqs.(9.3) and (9.4) lies in the boundary conditions Eqs.(9.2c) and (9.4c).

We therefore integrate the equations

$$\frac{dp}{-2c_5 p} = \frac{dq}{-3c_5 q} = da \tag{9.22}$$

from the variables in the boundary value problem, namely,

$$p(= \frac{df}{d\eta}) \;\; and \;\; q(= \frac{d^2 f}{d\eta^2}$$

to the variables in the initial value problem, namely,

$$p^*(= \frac{dg}{\partial \xi}) \;\; and \;\; q^*(= \frac{d^2 g}{d\xi^2})$$

The results are

$$\frac{q^*}{q} = e^{-3c_5 a} \;\; and \;\; \frac{p^*}{p} = e^{-2c_5 a} \tag{9.23a, b}$$

Evaluating Eq.(9.23a) at $\eta = \xi = 0$:

$$\frac{d^2 f(0)}{d\eta^2} = e^{-3c_5 a} \tag{9.24}$$

Evaluating Eq.(9.23b) at $\eta = \xi = \infty$:

$$\frac{dg(\infty)}{d\xi} = e^{-2c_5 a} \tag{9.25}$$

Eliminating $c_5 a$ from Eqs.(9.24) and (9.25), we obtain the relationship

$$\frac{d^2 f(0)}{d\eta^2} = (\frac{dg(\infty)}{d\xi})^{-3/2}$$

which is the same as Eq.(9.10).

9.2 Longitudinal Impact of Nonlinear Viscoplastic Rods

We now consider the problem of longitudinal impact of a semi-infinite nonlinear viscoplastic rod that is subjected to a constant velocity impact, v_0, as shown in Fig.9.2.

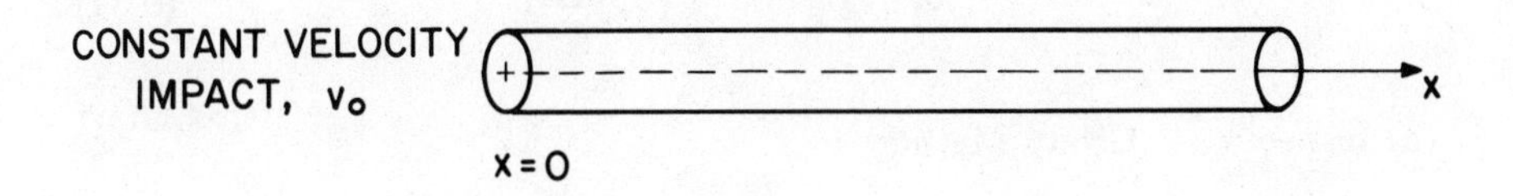

Fig. 9.2 One-Dimensional Viscoplastic Impact

The governing equation is quasilinear parabolic and can be written as[5]:

$$\gamma q\left(\frac{\partial^2 v}{\partial x^2}\right)\left(-\frac{\partial v}{\partial x}\right)^{q-1} = \frac{\partial v}{\partial t} \tag{9.26}$$

v is the particle velocity, and q is the nonlinear exponent of the constitutive relationship

$$\frac{\partial e}{\partial t} = D\left(\frac{\sigma}{\sigma_0} - 1\right)^{1/q}$$

x is the space coordinate, t is the time, σ is the stress and e is the strain. The constants γ, D and q depend on the material of the rod.

Using the similarity transformation

$$v = v_0 f(\eta)$$

where

$$\eta = \frac{kx}{t^m} \quad ; \quad k = \frac{1}{(\gamma q)^{1/(q+1)}\, v_0^{(q-1)/(q+1)}} \quad ; \quad m = \frac{1}{q+1} \tag{9.27}$$

Eq.(9.27) can be written as

$$\frac{d^2 f}{\partial \eta^2} - m\eta\left(-\frac{df}{d\eta}\right)^{2-q} = 0 \tag{9.28}$$

The boundary conditions $v(0,t) = v_0$ and $v(\infty,t) = 0$ transform to

$$f(0) = 1 \;\; ; \;\; f(\infty) = 0 \tag{9.29}$$

Making the change of variable $f = 1 - F$, Eqs.(9.28) and (9.29) becomes:

$$\frac{d^2F}{d\eta^2} + m\eta\left(\frac{dF}{d\eta}\right)^{2-q} = 0 \tag{9.30}$$

$$F(0) = 0 \;\; and \;\; F(\infty) = 1 \tag{9.31}$$

(a) Inspectional Group Method

We now define a linear group of transformation, G,

$$G \; : \;\; \eta = A^{\alpha_1}\xi \;\; ; \;\; F = A^{\alpha_2} g$$

so that the boundary value problem can be transformed to

$$\frac{d^2g}{d\xi^2} + m\xi\left(\frac{dg}{d\xi}\right)^{2-q} \tag{9.32}$$

$$g(0) = 0 \;\; and \;\; \frac{dg(0)}{d\xi} = 1 \tag{9.33}$$

Transforming Eq.(9.30) under the group G:

$$\frac{A^{\alpha_2}}{A^{2\alpha_1}}\frac{d^2g}{d\xi^2} + A^{\alpha_1(q-1)+\alpha_2(2-q)}\, m\xi\left(\frac{dg}{d\xi}\right)^{2-q} = 0$$

If we set

$$\alpha_2 - 2\alpha_1 = (q-1)\alpha_1 + (2-q)\alpha_2$$

then Eq.(9.32) is obtained. Therefore,

$$\frac{\alpha_1}{\alpha_2} = \frac{q-1}{q+1} \tag{9.34}$$

The boundary condition $F(0) = 0$ is transformed to $g(0) = 0$. However, the boundary condition $F(\infty)$ is transformed as follows:

$$F(\infty) = A^{\alpha_2} g(\infty) = 1$$

Therefore,

$$A = \left[\frac{1}{g(\infty)}\right]^{1/\alpha_2} \tag{9.35}$$

We now determine A such that "missing" boundary condition

$$\frac{dF(0)}{d\eta} = A \quad or \quad A^{\alpha_2 - \alpha_1} \frac{dg(0)}{d\xi} = A$$

Since $dg(0)/d\xi = 1$, we have:

$$\alpha_2 - \alpha_1 = 1 \tag{9.36}$$

From Eqs.(9.34) and (9.36),

$$\alpha_1 = \frac{1}{2}(q-1) \quad and \quad \alpha_2 = \frac{1}{2}(q+1) \tag{9.37}$$

The numerical procedure would proceed as follows:

(i) Solve Eqs.(9.32) and (9.33) as an initial value problem using forward integration. Obtain the value of $g(\infty)$.

(ii) Evaluate A using Eqs.(9.35) and (9.37).

(iii) Using the group of transformations, G, the variation of F vs. η can now be determined.

For different values of the nonlinear exponent, q, the gradient $F'(0) = A$ is compared with closed-form results[5] and the agreement is found to be good.

(b) Infinitesimal Group Method

Eq.(9.30) can be written as

$$k + m\eta p^{2-q} = 0 \tag{9.38}$$

where

$$p = \frac{dF}{d\eta} \quad ; \quad k = \frac{d^2F}{d\eta^2}$$

Introducing the infinitesimal group of transformations:

$$\bar{\eta} = \eta + \epsilon\phi(\eta, F) + O(\epsilon^2)$$

$$\bar{F} = F + \epsilon\theta(\eta, F) + O(\epsilon^2)$$

$$\bar{p} = p + \epsilon\varsigma(\eta, F, p) + O(\epsilon^2)$$

$$\bar{k} = k + \epsilon\delta(\eta, F, p, k) + O(\epsilon^2) \tag{9.39}$$

In terms of the characteristic function, W,

$$\phi = \frac{\partial W}{\partial p} \quad ; \quad \theta = p\frac{\partial W}{\partial p} - W$$

$$\varsigma = -X(W)$$

$$\delta = -[X^2 + 2kX\frac{\partial W}{\partial p} + k\frac{\partial W}{\partial F}] \tag{9.40}$$

with

$$X(\ ) \equiv \frac{\partial(\ )}{\partial p} + p\frac{\partial(\ )}{\partial F}$$

$$X^2(\ ) = \frac{\partial^2(\ )}{\partial \eta^2} + 2p\frac{\partial^2(\ )}{\partial \eta \partial F} + p^2\frac{\partial^2(\ )}{\partial F^2}$$

Again, the objective here is to transform the given boundary value problem to an initial value problem suitable for numerical forward integration procedures.

Invariance of Eq.(9.38) under the group defined by Eqs.(9.39) requires that

$$\delta + mp^{2-q}\phi + m(2-q)\eta p^{1-q}\varsigma = 0 \tag{9.41}$$

Substituting Eqs.(9.40) into Eq.(9.41), we get

$$-\frac{\partial^2 W}{\partial \eta^2} - 2p\frac{\partial^2 W}{\partial \eta \partial F} - p^2\frac{\partial^2 W}{\partial F^2} + m\eta p^{2-q}(2X\frac{\partial W}{\partial p} + \frac{\partial W}{\partial F})$$

$$+mp^{2-q}\frac{\partial W}{\partial p} - m(2-q)\eta p^{1-q}X = 0 \tag{9.42}$$

Since $W(\eta, F, p)$ is linear in p, we let

$$W(\eta, F, p) = W_1(\eta, F)p + W_2(\eta, F) \tag{9.43}$$

Substituting Eq.(9.43) into Eq.(9.42), and equating the coefficients to zero:

$$\frac{\partial^2 W_2}{\partial \eta^2} = 0$$

$$\frac{\partial^2 W_1}{\partial \eta^2} + 2\frac{\partial^2 W_2}{\partial \eta \partial F} = 0$$

$$2\frac{\partial^2 W_1}{\partial \eta \partial F} + \frac{\partial^2 W_2}{\partial F^2} = 0$$

$$\frac{\partial^2 W_1}{\partial F^2} = 0 \tag{9.44}$$

$$mp\eta\frac{\partial W_1}{\partial \eta} - m(1-q)\eta\frac{\partial W_2}{\partial F} + mW_1 = 0$$

$$\frac{\partial W_1}{\partial F} = 0$$

$$\frac{\partial W_2}{\partial \eta} = 0$$

The characteristic function can be obtained by solving Eqs. (9.44) in the following form:

$$W(\eta, F, p) = \left(\frac{1-q}{1+q} c_3 \eta\right) p + \left(c_3 F + c_4\right) \tag{9.45}$$

The infinitesimals ϕ, θ and ς can now be written as:

$$\phi = \frac{1-q}{1+q} c_3 \eta \;\; ; \;\; \theta = -c_3 F - c_4 \;\; ; \;\; \varsigma = -\frac{2c_3}{1+q} p \tag{9.46}$$

We now solve the system of equations

$$\frac{d\eta}{\phi} = \frac{dF}{\theta} = \frac{dp}{\varsigma} = da \tag{9.47}$$

Integrating Eqs.(9.47) so that the variables of the boundary value problem, Eqs.(9.30) and (9.31) can be related to those of the initial value problem, Eqs.(9.32) and (9.33), we obtain the following relationships.

$$\frac{\xi}{\eta} = e^{(1-q)/(1+q) c_3 a} \tag{9.48a}$$

$$\frac{g}{F} = e^{-c_3 a} \tag{9.48b}$$

$$\frac{p^*}{p} = e^{-2c_3 a/(1+q)} \tag{9.48c}$$

Evaluating Eq.(9.48b) at η approaches infinity, we get:

$$\frac{g(\infty)}{F(\infty)} = e^{-c_3 a}$$

Since $F(\infty) = 1$, we have

$$g(\infty) = e^{-c_3 a} \tag{9.49}$$

Evaluating Eq.(9.48c) at η equals to zero, we get

$$\frac{\frac{dg(0)}{d\xi}}{\frac{dF(0)}{d\eta}} = exp\left(-\frac{2c_3 a}{1+q}\right) \tag{9.50}$$

Since we have stipulated that

$$\frac{dg(0)}{d\xi} = 1$$

in the initial value problem, we get

$$\frac{dF(0)}{d\eta} = exp(\frac{2c_3 a}{1+q}) \tag{9.51}$$

Eliminating 'a' from Eqs.(9.49) and (9.51),we get

$$\frac{dF(0)}{d\eta} = [\frac{1}{g(\infty)}]^{2/(q+1)} \tag{9.52}$$

Eq.(9.52), which is obtained by the infinitesimal group method, is the same as Eq.(9.35) obtained by the inspectional group method.

9.3 Summary

Group-theoretic techniques for transforming a boundary value problem to an initial value problem were covered in this chapter. Two methods for obtaining such transformations were discussed: (1) the inspectional group method, and (2) the infinitesimal group method. The basic difference between the inspectional and the infinitesimal group methods is that in the latter, the transformation is deduced systematically by starting out with a general group of transformations. The required transformation that convert a boundary value problem to an equivalent initial value description were obtained by stipulating that: (a) the governing differential equations be invariant under the group of transformations, and (b) the parameter of transformation is identified as the "missing" boundary condition. For further details on the application of the group-theoretic techniques to boundary value problems, readers should refer to the text by Na[6].

REFERENCES

1. Toepfer, K.,"Grenzschichten in Flussigkeiten mit kleiner reibung",Z. Math. Phys.,60 (1912).
2. Klamkin,M.S., "On the Transformation of a Class of Boundary Value Problems into Initial Value Problems for Ordinary Differential Equations", SIAM Rev.,4, 43-47 (1962).
3. Na, T.Y.,"Transforming Boundary Conditions to Initial Conditions for Ordinary Differential Equations", SIAM Rev.,9, 204-210 (1967).
4. Goldstein, S., Modern Developments in fluid Mechanics, Oxford University Press, London, 1957.
5. Seshadri, R. and Na, T.Y. ,"Invariant Solution for Nonlinear Viscoplastic Impact",Ind. Math., 25, 37-44 (1975).
6. Na, T.Y.,Computational Methods in Eng. Boundary Value Problems, Academic Press (1979).

Chapter 10

FROM NONLINEAR TO LINEAR DIFFERENTIAL EQUATIONS USING TRANSFORMATION GROUPS

10.0 Introduction

The mathematical descriptions of large number of physical problems arising in science and engineering manifest themselves as nonlinear differential equations. Since there is an abundance of methods for dealing with linear differential equations, a popular practice has been to introduce some form of approximation that would linearize the nonlinear equation. These approximations usually impose certain restrictions on the solutions. In this chapter, we will discuss procedures for deriving mappings based on group-theoretic motivations that transform a nonlinear differential equation into a linear differential equation.

Na and Hansen[1] and Bluman[2] have proposed deductive procedures for deriving mappings that transform a nonlinear differential equation to a linear form. Their procedures are based on the use of Lie's group of "point transformations" which act on a finite dimensional space. Na and Seshadri[3] have proposed an approach that is based on simple groups of transformations. The underlying group-theoretic concept of invariance is used in all of the above procedures.

It has been recently shown that differential equations can be invariant under a continuous group of transformations beyond point or contact transformations[4]. These continuous group of transformations commonly known as "Noether transformations" or "Lie-Backlund (LB) transformation" act on an infinite dimensional space. The underlying basis of these procedures can be summarized as follows: Any linear differential equation which admits a nontrivial one-parameter point Lie group of transformations is invariant under an infinite number of one-parameter LB transformations through superposition. Moreover,every known nonlinear partial differential equation invariant under LB transformations can be associated with some corresponding linear partial differential equations. Readers are referred to the works of Anderson et al[5] and Kumei and Bluman[6] for details on the use of LB transformations for discovering mappings that transform nonlinear to linear differential equations.

In this chapter, procedures based on Lie's group of point transformations will be discussed in some detail. Effort has been made to keep the treatment of the subject as simple as possible, so that the novice can grasp the underlying principles involved.

10.1 From Nonlinear to Linear Differential Equations

The underlying principle in any group technique that transforms one differential equation to another can be stated as follows: "If a transformation maps any solution of a differential equation H_M into solution of a differential equation H_N, then it is necessary that this mapping transforms the Lie group of H_M into the Lie group of H_N". Therefore, let us then say that the differential equations H_M and H_N are *invariant* under groups G_M and G_N, respectively. If a transformation can be found that maps G_M into G_N, then the differential equation H_M is mapped into H_N.

Let us now consider a second order partial differential equation to represent H_M and H_N:

$$H_M : u_{xx} = H_M(x, t, u, u_x, u_t, u_{xt}, u_{tt}) \tag{10.1}$$

$$H_N : v_{xx} = H_N(x, t, v, v_x, v_t, v_{xt}, v_{tt}) \tag{10.2}$$

Introducing the notation

$$\varsigma_1 = x ; \ \varsigma_2 = t ; \ \varsigma_3 = u ; \ \varsigma_4 = u_x ; \ \varsigma_5 = u_t$$

$$\varsigma_6 = u_{xt} ; \ \varsigma_7 = u_{tt} ; \ \varsigma_8 = u_{xx},$$

Eq.(10.1) can be written as:

$$\varsigma_8 = H_M(\varsigma_1, \varsigma_2, \varsigma_3, \varsigma_4, \varsigma_5, \varsigma_6, \varsigma_7). \tag{10.3}$$

We now define an infinitesimal Lie group (G_M) as:

$$\bar{\varsigma_1} = \varsigma_1 + \epsilon S_M^{(1)}(\varsigma_1, \varsigma_2, \varsigma_3) + O(\epsilon^2) \tag{10.4a}$$

$$\bar{\varsigma_2} = \varsigma_2 + \epsilon S_M^{(2)}(\varsigma_1, \varsigma_2, \varsigma_3) + O(\epsilon^2) \tag{10.4b}$$

$$\bar{\varsigma_3} = \varsigma_3 + \epsilon S_M^{(3)}(\varsigma_1, \varsigma_2, \varsigma_3) + O(\epsilon^2) \tag{10.4c}$$

$$\bar{\varsigma_k} = \varsigma_k + \epsilon S_M^{(k)}(\varsigma_1, \varsigma_2, \varsigma_3) + O(\epsilon^2) \tag{10.4d}$$

where k=4,5,6,7 and 8, and ϵ is the parameter of the infinitesimal group. We can write Eq.(10.4) in an abbreviated form as

$$G_M(S_M^{(1)}, S_M^{(2)}, S_M^{(3)}; \epsilon) \tag{10.5}$$

Similarly, the group G_N can be written as:

$$G_N(S_N^{(1)}, S_N^{(2)}, S_N^{(3)}; \epsilon) \tag{10.6}$$

where S_M and S_N are the infinitesimals and their extensions.

If there exists a mapping (T) such that *

$$T: \; v = F(\varsigma_1, \varsigma_2, \varsigma_3, \varsigma_4, \varsigma_5) \tag{10.7}$$

maps the solution of H_M into those of H_N, then F must necessarily satisfy the following relationship arising from invariance of Eq.(10.7) under Eq.(10.4) to (10.6):

$$S_N^{(3)} = \sum_{i=1}^{5} S_M^{(i)} \frac{\partial F}{\partial \varsigma_i} \tag{10.8}$$

If we seek a transformation of dependent variables only, then

$$S_M^{(1)} = S_N^{(1)} \quad ; \quad S_M^{(2)} = S_N^{(2)} \tag{10.9}$$

Eqs.(10.8) and (10.9) express the underlying concept of invariance, and relate equations H_M and H_N.

Two methods for deducing the required mapping that would transform a nonlinear differential equation into a linear differential equation is now discussed.

(i) The Inspectional Group Method

In an effort to enable a non-specialist in group theory to use these powerful concepts in a quick and simple manner, a procedure entitled "inspectional group method" was proposed [3]. For the purpose of discussion, we consider a one-parameter linear group. However, other groups such as spiral groups etc. can be used.

The key steps are as follows:

(1) Define the *assumed* group of transformations as

$$G_M: \; (\bar{x} = a^m x \; ; \; \bar{t} = a^n t \; ; \; \bar{u} = a^p u) \tag{10.10a}$$

$$G_N: \; (\bar{x} = a^m x \; ; \; \bar{t} = a^n t \; ; \; \bar{v} = a^q v) \tag{10.10b}$$

(2) Invoke invariance of the differential equations (10.1) and (10.2) under groups (10.10) , respectively, and determine the relationship between p, m and n, and the relationship between q, m and n.

(3) Consistent with the "assumed" group,a mapping of the following form is sought:

$$T: \; u = C x^{\alpha_1} t^{\alpha_2} v^{\alpha_3} v_x{}^{\alpha_4} v_t{}^{\alpha_5} \tag{10.11}$$

The constants α_1 to α_5 are determined so that

$$T: \; \bar{u} = C \bar{x}^{\alpha_1} \bar{t}^{\alpha_2} \bar{v}^{\alpha_3} \bar{v}_{\bar{x}}^{\alpha_4} \bar{v}_{\bar{t}}^{\alpha_5} \tag{10.12}$$

* The necessary and sufficient condition for the existence of a one to one transformation of a system of nonlinear differential equations to a system of linear equations is discussed by Kumei and Bluman [6].

In practice, some trial and error is involved at this stage. We can assume a trial form of Eq.(10.11) by setting some of the $\alpha' s$ equal to zero. The only exception is when

$$\alpha_3 = \alpha_4 = \alpha_5 = 0$$

For this case, the mapping is invalid.

(4) Substitute the trial mapping (T) into equation H_M and check if equation H_N is obtained. If it does, the value of C can be readily obtained. If it fails to yield equation H_N, then proceed to find another mapping using different $\alpha' s$.

While the inspectional group method has been described by using a one-parameter linear group of transformations, any other inspectional groups such as spiral groups can be used. The reader should try out all possible groups on order to come up with the required nonlinear to linear mapping. Limitations of the inspectional group method is discussed in section 10.4.

(ii) The Infinitesimal Group Method

The infinitesimal group method is mathematically more rigorous and complete. The key steps for this method are as follows:

(1) For the given nonlinear equation H_M, and the desired linear form H_N, the respective infinitesimals or transformation functions

$$\left(S_M^{(1)}, S_M^{(2)}, S_M^{(3)}\right) \ \ and \ \ \left(S_N^{(1)}, S_N^{(2)}, S_N^{(3)}\right)$$

are obtained by determining the "characteristic function", W, as described in section 3.2.

(2) Since only transformation of dependent variables is sought, it is stipulated that

$$S_M^{(1)} = S_N^{(1)} \ \ ; \ \ S_M^{(2)} = S_N^{(2)}$$

(3) The required nonlinear to linear transformation (T),

$$T: \ v = F(\varsigma_1, \varsigma_2, \varsigma_3, \varsigma_4, \varsigma_5)$$

is obtained by solving the subsystem

$$S_N^{(3)} = \sum_{i=1}^{5} S_M^{(i)} \frac{\partial F}{\partial \varsigma_i}$$

or

$$\frac{dF}{S_N^{(3)}} = \frac{d\varsigma_1}{S_M^{(1)}} = \frac{d\varsigma_2}{S_M^{(2)}} = \ldots\ldots = \frac{d\varsigma_5}{S_M^{(5)}}$$

10.2 Application to Ordinary Differential Equations -Bernoulli's Equation

Consider, as an example, the first order nonlinear ordinary differential equation [7]:

$$H_M : \frac{du}{dx} + B(x)u = R(x)u^k \tag{10.13}$$

The objective here is to systematically discover a mapping (T) using the inspectional as well as infinitesimal group methods that will transform Eq.(10.13) into the linear form:

$$H_N : \frac{dv}{dx} + C(x)v = D(x) \tag{10.14}$$

(a) Inspectional Group Method:

We now define a group of transformations G_M and G_N as follows:

$$G_M : \left(u = a^{p_1}\bar{u} \ ; \ x = a^r \bar{x}\right) \tag{10.15a}$$

$$G_N : \left(v = a^{p_2}\bar{v} \ ; \ x = a^r \bar{x}\right) \tag{10.15b}$$

For Eqs.(10.13) and (10.14) to be invariant under G_M and G_N respectively, their coefficients have to be expressible in the following form:

$$B(x) = b_0 x^{m_1} \ ; \ R(x) = r_0 x^{n_1} \tag{10.16a}$$

$$C(x) = c_0 x^{m_2} \ ; \ D(x) = d_0 x^{n_2} \tag{10.16b}$$

These types of restrictions are typical of the inspectional methods where assumed transformation groups are used.

Invariance of Eq.(10.13) under group G_M gives the following relationship:

$$p_1(1-k) = r(n_1 - m_1) \tag{10.17}$$

Similarly, invariance of Eq.(10.14) under group G_N leads to:

$$p_2 = r(n_2 - m_2) \tag{10.18}$$

We now seek a trial mapping (T)

$$T: \ u = \lambda v^{\delta_1} x^{\delta_2} \tag{10.19}$$

where δ_1 and δ_2 are as yet to be determined. Invariance of Eq.(10.19) under G_M and G_N along with the use of Eqs.(10.17) and (10.18) gives

$$\delta_1 = \left(\frac{n_1 - m_1}{n_2 - m_2}\right)\frac{1}{1-k} - \frac{1}{n_2 - m_2} \tag{10.20}$$

The Leibnitz transformation is obtained when $m_1 = m_2, n_1 = n_2$ and $\delta_2 = 0$. Therefore,

$$\delta_1 = \frac{1}{1-k} \tag{10.21}$$

where λ can be found to equal to one, by substituting Eq.(10.21) into Eq.(10.13) and then comparing with Eq.(10.14).

It is seen that that by assuming other forms of "trial mappings", T,the linear differential equation will not result. The trial and error process involved is certainly a disadvantage of the inspectional group method. However, whenever applicable, the method is very simple to apply. The limitations of this method are discussed in section 10.4.

(b) Infinitesimal Group Method:

We will now rewrite the Bernoulli's equation as:

$$H_M : \ p + B(x)u = R(x)u^k \tag{10.13a}$$

where $p = du/dx$.

Similarly, the linear version can be written as

$$H_N : \ q + C(x)v = D(x) \tag{10.14a}$$

where $q = dv/dx$.

An infinitesimal group of transformations is defined as follows:

$$\begin{aligned} G_m : \bar{x} &= x + \epsilon\xi(x,u,p) + O(\epsilon^2) \\ \bar{u} &= u + \epsilon\theta(x,u,p) + O(\epsilon^2) \\ \bar{p} &= p + \epsilon\pi(x,u,p) + O(\epsilon^2) \end{aligned} \tag{10.22a}$$

and

$$\begin{aligned} G_N : \bar{x} &= x + \epsilon\varsigma(x,v,q) + O(\epsilon^2) \\ \bar{v} &= v + \epsilon S(x,v,q) + O(\epsilon^2) \\ \bar{q} &= q + \epsilon Q(x,v,q) + O(\epsilon^2) \end{aligned} \tag{10.22b}$$

In terms of the characteristic functiion, W_1,

$$\begin{aligned} \xi &= \frac{\partial W_1}{\partial p} \\ \theta &= p\frac{\partial W_1}{\partial p} - W_1 \\ \pi &= -\frac{\partial W_1}{\partial x} - p\frac{\partial W_1}{\partial u} \end{aligned} \tag{10.23}$$

Similarly, in terms of the characteristic function, W_2,

$$\varsigma = \frac{\partial W_2}{\partial q}$$

$$S = q\frac{\partial W_2}{\partial q} - W_2 \tag{10.24}$$

$$Q = -\frac{\partial W_2}{\partial x} - q\frac{\partial W_2}{\partial v}$$

If the variable x remains unchanged after the transformations, then

$$\xi = 0 \;\; or \;\; \frac{\partial W_1}{\partial p} = 0 \tag{10.25a}$$

$$\varsigma = 0 \;\; or \;\; \frac{\partial W_2}{\partial q} = 0 \tag{10.25b}$$

This stipulation is more restrictive than the requirement that ξ and ς should be equal, but non-zero, for a nonlinear to linear transformation to exist. Rewriting Eq.(10.13a) as

$$F_1 \equiv p + B(x)u - R(x)u^k = 0 \tag{10.13b}$$

we find that Eq.(10.13b) is invariant under G_M if

$$\theta[\frac{\partial F_1}{\partial u} + \pi(\frac{\partial F_1}{\partial p})] = 0 \tag{10.26}$$

Substituting Eq.(10.23) into Eq.(10.26),

$$(B - kRu^{k-1})(p\frac{\partial W_1}{\partial p} - W_1) - (\frac{\partial W_1}{\partial x} + p\frac{\partial W_1}{\partial u}) = 0 \tag{10.27}$$

Eq.(10.27) is a first order linear partial differential equation in W_1. Since $\partial W_1/\partial p = 0$, Eq.(10.27) can be written as:

$$(B - kRu^{k-1})\, W_1 + \frac{\partial W_1}{\partial x} + p\frac{\partial W_1}{\partial u} = 0 \tag{10.28}$$

Eq.(10.28) is linear in W_1, therefore we can separate the variables x and u as:

$$W_1(x,u) = \phi_1(x)\psi(u) \tag{10.29}$$

Substituting W_1 from Eq.(10.29) into Eq.(10.28) , we get

$$(B - kRu^{k-1}) + \frac{1}{\phi_1}\frac{\partial \phi_1}{\partial x} + p\frac{1}{\psi_1}\frac{\partial \psi_1}{\partial u} \tag{10.30}$$

Introducing p from Eq.(10.13b) into Eq.(10.30) and rearranging, the following equation can be obtained:

$$[(1-k)B+\frac{1}{\phi_1}\frac{\partial\phi_1}{\partial x}]-p[\frac{k}{u}+\frac{1}{\psi_1}\frac{\partial\psi_1}{\partial u}]=0 \tag{10.31}$$

from which

$$\frac{\partial\phi_1}{\partial x}=-(1-x)B\phi \tag{10.32}$$

$$\frac{\partial\psi_1}{\partial u}=\frac{k\psi}{u} \tag{10.33}$$

Solving Eq.(10.32) and Eq.(10.33), we get

$$\phi_1=exp(-\int(1-k)Bdu) \quad ; \quad \psi_1=u^k$$

Therefore, the characteristic function is

$$W_1(x,u)=u^k exp(-\int(1-k)Bdu) \tag{10.34}$$

Using Eq.(10.23) and (10.34), we have

$$\theta=p\frac{\partial W_1}{\partial p}-W_1=-u^k exp(-\int(1-k)Bdu) \tag{10.35}$$

In a similar fashion, we will proceed to analyze Eq.(10.14a). Here again, it is stipulated that the independent variable does not transform under G_N. Therefore,

$$\varsigma=\frac{\partial W_2}{\partial q}=0$$

Rewriting Eq.(10.14a) as

$$F_2\equiv q+C(x)-D(x)=0 \tag{10.14b}$$

the condition for invariance of the above equation under G_N can be written as

$$S\frac{\partial F_2}{\partial v}+Q\frac{\partial F_2}{\partial q}=0 \tag{10.36}$$

Substituting Eq.(10.24) into Eq.(10.36), we get

$$W_2C(x)+[\frac{\partial W_2}{\partial x}+q\frac{\partial W_2}{\partial v}]=0 \tag{10.37}$$

Since Eq.(10.37) is linear in W_2, we can assume that

$$W_2(x,v)=\phi_2(x)\psi_2(v) \tag{10.38}$$

Substituting Eq.(10.38) into Eq.(10.37), we get:

$$(C + \frac{1}{\psi_2}\frac{\partial \psi_2}{\partial v}) + q(\frac{1}{\phi_2}\frac{\partial \phi_2}{\partial x}) \tag{10.39}$$

Therefore,

$$\frac{1}{\phi_2}\frac{\partial \phi_2}{\partial x} = 0 \tag{10.40}$$

and

$$C + \frac{1}{\psi_2}\frac{\partial \psi_2}{\partial v} = 0 \tag{10.41}$$

The characteristic function, W_2, can be written as

$$W_2(x, v) = \alpha_0 \beta_0 exp(- \int C(x)\, dx) \tag{10.42}$$

Using the relationship defined in Eq.(10.24) the infinitesimal for v can be written as

$$S = q\frac{\partial W_2}{\partial q} - W_2 = -\alpha_0 \beta_0 exp(- \int C(x)\, dx) \tag{10.43}$$

To obtain the nonlinear to linear transformation

$$T: \quad u = u(v) \tag{10.44}$$

we solve the infinitesimal version of Eq.(10.44), which can be written as:

$$\theta = S\frac{du}{dv} \tag{10.45}$$

Therefore, using values of θ and S from Eq.(10.35) and Eq.(10.43),we get

$$u^k exp(- \int (1-k)B\, dx) = \alpha_0 \beta_0 exp[- \int C(x)\, dx]\frac{du}{dv} \tag{10.46}$$

Letting $C(x) = (1-k)B(x)$, we have

$$\frac{du}{u^k} = \frac{dy}{\alpha_0 \beta_0} \tag{10.47}$$

Integrating Eq.(10.47), we get

$$T: \quad v = \frac{\alpha_0 \beta_0}{1-k} u^{1-k}$$

When $\alpha_0\beta_0 = 1 - k$,the transformation (T) would map the nonlinear equation in the linear form.

Therefore, the required nonlinear to linear transformation (T) is

$$v = u^{1-k} \tag{10.48}$$

It can be seen that in both the inspectional and infinitesimal group methods, the underlying concept of invariance leads to the required mapping. While $B(x), C(x), D(x)$ and $R(x)$ have to be expressed in a power form in the inspectional group method because of the assumed group of transformations, no such restrictions are required in the infinitesimal group method.

10.3 Application to Partial Differential Equations-A Nonlinear Chemical Exchange Process

Consider now the nonlinear partial differential equation [7]

$$H_M : \; u_{xt} + u_t + u_x + u_x u_t = 0 \tag{10.49}$$

The above equation arises in a chemical exchange process between a solid bed and a fluid flowing through it, sediment transport in rivers, and in chromatography. We will now make a systematic attempt to obtain a mapping that will linearize Eq.(10.49) into the following form:

$$H_N \; : \quad v_{xt} + v_t + v_x = 0 \tag{10.50}$$

(a) Inspectional Group Method:

Eq.(10.49) is invariant under a group of transformations, G_M, defined as

$$G_M : \; \bar{u} = u + c_1 a \; ; \; \bar{x} = x \; ; \; \bar{t} = t \tag{10.51}$$

where a is the parameter of transformation.

By inspection, it can be seen that Eq.(10.50) is invariant under G_N,

$$G_N \; : \; \bar{v} = v e^{c_2 a} \; ; \; \bar{x} = x \; ; \; \bar{t} = t \tag{10.52}$$

Examination of groups G_M and G_N would reveal that a trial nonlinear to linear transformation can be expressed as

$$T : \quad v = \alpha e^{\beta u} \tag{10.53}$$

where α and β are to be determined such that Eq.(10.53) is invariant under G_M and G_N. Therefore, substituting Eq.(10.51) and (10.52) into Eq.(10.53)

$$e^{-c_2 a}\bar{v} = \alpha e^{\beta(\bar{u} - c_1 a)}$$

or,

$$e^{-c_2 a+\beta c_1 a}(\bar{v}) = \alpha e^{\beta \bar{u}}$$

For invariance of Eq.(10.53),

$$-c_2 a + \beta c_1 a = 0$$

or,

$$c_2 = \beta c_1 \tag{10.54}$$

The "trial" transformation can now be expressed as:

$$u = \frac{1}{\beta} ln(\frac{v}{\alpha}) \tag{10.55}$$

Substituting the transformation (10.55) into Eq.(10.49), we obtain the following:

$$\frac{\partial u}{\partial x} = \frac{1}{\beta}\frac{\alpha}{v}\frac{\partial v}{\partial x}$$

$$\frac{\partial u}{\partial t} = \frac{1}{\beta}\frac{\alpha}{v}\frac{\partial v}{\partial t} \tag{10.56}$$

$$\frac{\partial}{\partial t}(\frac{\partial u}{\partial x}) = \frac{\alpha}{\beta}[\frac{\partial}{\partial t}(\frac{1}{v}\frac{\partial v}{\partial x})] = \frac{\alpha}{\beta}[-\frac{1}{v^2}\frac{\partial v}{\partial x}\frac{\partial v}{\partial t} + \frac{1}{v}\frac{\partial^2 v}{\partial x \partial t}]$$

Therefore, Eq.(10.49) can be rewritten as:

$$\frac{\alpha}{\beta}[\frac{1}{v}\frac{\partial^2 v}{\partial x \partial t} - \frac{1}{v^2}\frac{\partial v}{\partial x}\frac{\partial v}{\partial t}] + \frac{\alpha}{\beta}\frac{1}{v}\frac{\partial v}{\partial t}$$

$$+\frac{\alpha}{\beta}\frac{1}{v}\frac{\partial v}{\partial x} + \frac{\alpha^2}{\beta^2 v^2}\frac{\partial v}{\partial x}\frac{\partial v}{\partial t} = 0 \tag{10.57}$$

If $\alpha = \beta = 1$, then Eq.(10.57) becomes:

$$\frac{\partial^2 v}{\partial x \partial t} + \frac{\partial v}{\partial x} + \frac{\partial v}{\partial t} = 0$$

which is the required linear form.

Therefore, the transformation (T) which maps given nonlinear equation into a linear form is

$$u = ln(v) \tag{10.58}$$

This is an example where the invariance is invoked under a spiral group of transformation.

(b) Infinitesimal Group Method:

The infinitesimal group method will now be used to discover a transformation (T) that will transform

$$H_M: \quad u_{xt} + u_t + u_x + u_x u_t = 0 \tag{10.49}$$

into a linear form

$$H_N: \quad v_{xt} + v_t + v_x = 0 \tag{10.50}$$

Defining an infinitesimal group, G_N:

$$\begin{aligned} G_N: \quad \bar{t} &= t + \epsilon T(t,x) + O(\epsilon^2) \\ \bar{x} &= x + \epsilon X(t,x) + O(\epsilon^2) \\ \bar{v} &= v + \epsilon V(t,x,u) + O(\epsilon^2) \end{aligned} \tag{10.59a}$$

where the infinitesimals are defined in section 2.9 as:

$$T = \frac{\partial W}{\partial p} \quad ; \quad X = \frac{\partial W}{\partial q} \quad ; \quad V = p\frac{\partial W}{\partial p} + q\frac{\partial W}{\partial q} - W$$

$$p = \frac{\partial v}{\partial t} \quad ; \quad q = \frac{\partial v}{\partial x}$$

The group is now extended as follows:

$$\begin{aligned} \bar{p} &= p + \epsilon\pi_1(t,x,v,p,q) + O(\epsilon^2) \\ \bar{q} &= q + \epsilon\pi_2(t,x,v,p,q) + O(\epsilon^2) \\ \bar{r} &= r + \epsilon\phi_1(t,x,v,p,q,s,f) + O(\epsilon^2) \\ \bar{s} &= s + \epsilon\phi_2(t,x,v,p,q,r,s,f) + O(\epsilon^2) \\ \bar{f} &= f + \epsilon\phi_3(t,x,v,p,q,s,f) + O(\epsilon^2) \end{aligned} \tag{10.59b}$$

where

$$r = \frac{\partial^2 v}{\partial t^2} \quad ; \quad s = \frac{\partial^2 v}{\partial t \partial x} \quad and \quad f = \frac{\partial^2 v}{\partial x^2}$$

and

$$\pi_1 = -\frac{\partial W}{\partial t} - p\frac{\partial W}{\partial v} \quad ; \quad \pi_2 = -\frac{\partial W}{\partial x} - q\frac{\partial W}{\partial v}$$

$$\phi_2 = -\frac{\partial^2 W}{\partial x \partial t} - p\frac{\partial^2 W}{\partial x \partial v} - q\frac{\partial^2 W}{\partial v \partial t} - r\frac{\partial^2 W}{\partial x \partial p}$$

$$-s\left(\frac{\partial W}{\partial v} + \frac{\partial^2 W}{\partial x \partial q} + \frac{\partial^2 W}{\partial t \partial p}\right) - f\left(\frac{\partial^2 W}{\partial t \partial q}\right) + p^2\frac{\partial^3 W}{\partial x \partial v \partial p}$$

$$+pq\left(\frac{\partial^3 W}{\partial x \partial v \partial q} + \frac{\partial^3 W}{\partial v \partial t \partial p} - \frac{\partial^2 W}{\partial v^2}\right) + q^2\frac{\partial^3 W}{\partial v \partial t \partial q}$$

$$+p^2 q\frac{\partial^3 W}{\partial p\partial v^2}+pq^2\frac{\partial^3 W}{\partial q\partial v^2}+ps\frac{\partial^2 W}{\partial v\partial p}+qs\frac{\partial^2 W}{\partial v\partial q}$$

Eq.(10.50) can be rewritten as

$$H_N(v)\ :\quad s+p+q=0 \tag{10.50a}$$

The general invariance relationship is given by

$$T\frac{\partial H_N}{\partial t}+X\frac{\partial H_N}{\partial x}+V\frac{\partial H_N}{\partial v}+\pi_1\frac{\partial H_N}{\partial p}+\pi_2\frac{\partial H_N}{\partial q}$$

$$+\phi_1\frac{\partial H_N}{\partial r}+\phi_2\frac{\partial H_N}{\partial s}+\phi_3\frac{\partial H_N}{\partial t}=0 \tag{10.60}$$

Substituting Eq.(10.50) into Eq.(10.60), we get:

$$\pi_1+\pi_2+\phi_2=0 \tag{10.61}$$

In terms of the characteristic function, W, Eq.(10.61), after replacing s by $-(p+q)$ can be expressed as:

$$\frac{\partial W}{\partial t}+p\frac{\partial W}{\partial v}+\frac{\partial W}{\partial x}+q\frac{\partial W}{\partial v}+\frac{\partial^2 W}{\partial x\partial t}$$

$$+p\frac{\partial^2 W}{\partial x\partial v}+q\frac{\partial^2 W}{\partial v\partial t}+r\frac{\partial^2 W}{\partial x\partial p}-(p+q)(\frac{\partial W}{\partial v}+\frac{\partial^2 W}{\partial x\partial q}$$

$$+\frac{\partial^2 W}{\partial p\partial t}-p\frac{\partial^2 W}{\partial v\partial p}-q\frac{\partial^2 W}{\partial v\partial q}+f\frac{\partial^2 W}{\partial t\partial q}$$

$$-p^2\frac{\partial^3 W}{\partial x\partial v\partial p}-pq(\frac{\partial^3 W}{\partial x\partial v\partial q}+\frac{\partial^3 W}{\partial v\partial t\partial p}-\frac{\partial^2 W}{\partial v^2})$$

$$-q^2\frac{\partial^3 W}{\partial v\partial t\partial q}-p^2 q\frac{\partial^3 W}{\partial v^2\partial p}-pq\frac{\partial^3 W}{\partial v^2\partial q}=0 \tag{10.62}$$

Since W is linear in both p and q,

$$W(t,x,v,p,q)\ =\ W_1(t,x,u)p+W_2(t,x,v)q+W_3(t,x,v). \tag{10.63}$$

Substituting Eqs.(10.63) into (10.62), we get

$$(\frac{\partial W_3}{\partial t}+\frac{\partial W_3}{\partial x}+\frac{\partial^2 W_3}{\partial x\partial t})+(\frac{\partial W_1}{\partial x}+\frac{\partial^2 W_1}{\partial x\partial t}+\frac{\partial^2 W_3}{\partial x\partial v}-\frac{\partial W_2}{\partial x})p$$

$$+(\frac{\partial W_2}{\partial t}+\frac{\partial^2 W_2}{\partial x\partial t}+\frac{\partial^3 W_3}{\partial v\partial t}-\frac{\partial W_1}{\partial t})q+\frac{\partial W_1}{\partial v}p^2+\frac{\partial^2 W_2}{\partial v^2}q^2$$

$$+(\frac{\partial W_1}{\partial v}+\frac{\partial W_2}{\partial v}+\frac{\partial^2 W_3}{\partial v^2})pq+\frac{\partial W_1}{\partial x}r+\frac{\partial W_2}{\partial t}f=0$$

Equating the coefficients of $p^0, p, q, p^2, q^2, pq, r$ and f, we get:

$$\frac{\partial W_3}{\partial t} + \frac{\partial W_3}{\partial x} + \frac{\partial^2 W_3}{\partial x \partial t} = 0 \tag{10.64a}$$

$$\frac{\partial W_1}{\partial t} + \frac{\partial^2 W_1}{\partial x \partial t} + \frac{\partial^2 W_3}{\partial v \partial t} - \frac{\partial W_2}{\partial x} = 0 \tag{10.64b}$$

$$\frac{\partial W_2}{\partial t} + \frac{\partial^2 W_2}{\partial x \partial t} + \frac{\partial^2 W_3}{\partial v \partial t} - \frac{\partial W_1}{\partial t} = 0 \tag{10.64c}$$

$$\frac{\partial W_1}{\partial v} = 0 \tag{10.64d}$$

$$\frac{\partial W_2}{\partial v} = 0 \tag{10.64e}$$

$$\frac{\partial W_1}{\partial v} + \frac{\partial W_2}{\partial v} + \frac{\partial^2 W_3}{\partial v^2} = 0 \tag{10.64f}$$

$$\frac{\partial W_1}{\partial x} = 0 \tag{10.64g}$$

$$\frac{\partial W_2}{\partial t} = 0 \tag{10.64h}$$

Eqs.(10.64d) and (10.64g) show that W_1 is a function of t. Eqs.(10.64e) and (10.64h) show that $W_2 = W_2(x)$. Therefore, Eq.(10.64f) becomes

$$\frac{\partial^2 W_3}{\partial v^2} = 0 \tag{10.65}$$

so that

$$W_3(x,t,v) = W_{31}(t,x)v + W_{32}(t,x) \tag{10.65}$$

The simplified forms of Eqs.(10.64a) to (10.64c) can be written as:

$$\frac{\partial W_3}{\partial t} + \frac{\partial W_3}{\partial x} + \frac{\partial^2 W_3}{\partial x \partial t} = 0 \tag{10.66a}$$

$$\frac{\partial^2 W_3}{\partial x \partial v} - \frac{\partial W_2}{\partial x} = 0 \tag{10.66b}$$

$$\frac{\partial^2 W_3}{\partial v \partial t} - \frac{\partial W_1}{\partial t} = 0 \tag{10.66c}$$

Using Eq.(10.65) in the above equations, the following can be obtained:

$$\frac{\partial W_{31}}{\partial t} + \frac{\partial W_{31}}{\partial x} + \frac{\partial^2 W_{31}}{\partial x \partial t} = 0 \tag{10.67a}$$

$$\frac{\partial W_{32}}{\partial t} + \frac{\partial W_{32}}{\partial x} + \frac{\partial^2 W_{32}}{\partial x \partial t} = 0 \tag{10.67b}$$

$$\frac{\partial W_{31}}{\partial t} - \frac{\partial W_2}{\partial x} = 0 \tag{10.67c}$$

$$-\frac{\partial W_{31}}{\partial t} - \frac{\partial W_1}{\partial x} = 0 \tag{10.67d}$$

From Eq.(10.67c), we get

$$W_{31}(t,x) = W_2(x) + c_1(t) \tag{10.68}$$

From Eq.(10.67d), we get

$$W_{31}(t,x) = W_1(t) + c_2(x) \tag{10.69}$$

Comparing Eqs.(10.68) and (10.69), we conclude that

$$W_{31}(t,x) = W_1(t) + W_2(x) + c_3 \tag{10.70}$$

Substituting W_{31} from Eq.(10.70) into Eq.(10.67a), we get

$$\frac{\partial W_1}{\partial t} + \frac{\partial W_2}{\partial x} = 0$$

Therefore, we get

$$W_2(x) = c_4 x + c_5 \quad ; \quad W_1(t) = -c_4 t + c_6$$

Eq.(10.70) gives:

$$W_{31}(t,x) = c_4(x-t) + c_7 \tag{10.71}$$

The characteristic function, W, can therefore be written as:

$$\begin{aligned} W(t,x,v,p,q) &= (c_6 - c_4 t)p + (c_5 + c_4 x)q \\ &+ [c_7 + c_4(x-t)]v + W_{32}(t,x) \end{aligned} \tag{10.72}$$

where $W_{32}(t,x)$ is a function that satisfies Eq.(10.67b). By definition, Eq.(10.59a), we write

$$\begin{aligned} T &= \frac{\partial W}{\partial p} = c_6 - c_4 t \\ X &= \frac{\partial W}{\partial q} = c_5 + c_4 x \end{aligned} \tag{10.73}$$

$$V = p\frac{\partial W}{\partial p} + q\frac{\partial W}{\partial q} - W = [c_4(t-x) - c_7]v + W_{32}(t,x)$$

For the nonlinear equation, H_M,

$$H_M : \; u_{xt} + u_t + u_x + u_x u_t = 0 \tag{10.49}$$

the same procedure as for H_N is followed:

$$
\begin{aligned}
G_M \;:\; \bar{t} &= t + \epsilon\tau(t,x,u) + O(\epsilon^2) \\
\bar{x} &= x + \epsilon\psi(t,x,u) + O(\epsilon^2) \\
\bar{u} &= u + \epsilon U(t,x,u) + (\epsilon^2)
\end{aligned}
\tag{10.74}
$$

where

$$\tau = \frac{\partial W^*}{\partial p^*} \;\; ; \;\; \psi = \frac{\partial W^*}{\partial q^*}$$

$$U = p^* \frac{\partial W^*}{\partial p^*} + q^* \frac{\partial W^*}{\partial q^*} - W^*$$

with

$$p^* = \frac{\partial u}{\partial t} \;\; ; \;\; q^* = \frac{\partial u}{\partial x}$$

The group is extended as:

$$
\begin{aligned}
\bar{p}^* &= p^* + \epsilon\pi_1{}^*(t,x,u,p^*,q^*) + O(\epsilon^2) \\
\bar{q}^* &= q^* + \epsilon\pi_2{}^*(t,x,u,p^*,q^*) + O(\epsilon^2) \\
\bar{r}^* &= r^* + \epsilon\phi_1{}^*(t,x,u,p^*,q^*,r^*,s^*,f^*) + O(\epsilon^2) \\
\bar{s}^* &= s^* + \epsilon\phi_2{}^*(t,x,u,p^*,q^*,r^*,s^*,f^*) + O(\epsilon^2) \\
\bar{f}^* &= f^* + \epsilon\phi_3^*(t,x,u,p^*,q^*,r^*,s^*,f^*) + O(\epsilon^2)
\end{aligned}
\tag{10.75}
$$

with

$$r^* = \frac{\partial^2 u}{\partial t^2} \;\; ; \;\; s^* = \frac{\partial^2 u}{\partial t \partial x} \;\; ; \;\; f^* = \frac{\partial^2 u}{\partial x^2}$$

where $\pi_1^*\, and \pi_2^*$ can be expressed in terms of the characteristic function, W^*, in the same form as Eqs.(10.59b). Eq.(10.49) can be rewritten as

$$H_M : \; s^* + p^* + q^* + p^* q^* = 0 \tag{10.76}$$

The invariance relationship can be written as

$$
\begin{aligned}
&\tau \frac{\partial H_M}{\partial t} + \psi \frac{\partial H_M}{\partial x} + U \frac{\partial H_M}{\partial u} + \pi_1^* \frac{\partial H_M}{\partial p^*} + \pi_2^* \frac{\partial H_M}{\partial q^*} \\
&+ \phi_1^* \frac{\partial H_M}{\partial r^*} + \phi_2^* \frac{\partial H_M}{\partial s^*} + \phi_3^* \frac{\partial H_M}{\partial f^*} = 0
\end{aligned}
\tag{10.77}
$$

Substituting Eq.(10.76) into Eq.(10.77), we get

$$\pi_1^* + \pi_2^* + \phi_2^* + p^* \pi_2^* + q^* \pi_1^* = 0 \tag{10.78}$$

In the same fashion as the linear equation, the following equations can be obtained:

$$\frac{\partial W_3^*}{\partial t} + \frac{\partial W_3^*}{\partial x} + \frac{\partial^2 W_3^*}{\partial x \partial t} = 0 \tag{10.79a}$$

$$\frac{\partial W_1^*}{\partial x} + \frac{\partial^2 W_1^*}{\partial x \partial t} + \frac{\partial^2 W_3}{\partial x \partial u} - \frac{\partial W_2^*}{\partial x} + \frac{\partial W_3^*}{\partial x} = 0 \tag{10.79b}$$

$$\frac{\partial W_2^*}{\partial t} + \frac{\partial^2 W_2^*}{\partial x \partial t} + \frac{\partial^2 W_3^*}{\partial u \partial t} - \frac{\partial W_1^*}{\partial t} + \frac{\partial W_3^*}{\partial t} = 0 \tag{10.79c}$$

$$\frac{\partial W_1^*}{\partial u} + \frac{\partial W_1^*}{\partial t} = 0 \tag{10.79d}$$

$$\frac{\partial W_2^*}{\partial u} + \frac{\partial W_2^*}{\partial t} = 0 \tag{10.79e}$$

$$\frac{\partial W_1^*}{\partial u} + \frac{\partial W_2^*}{\partial u} + \frac{\partial^2 W_3^*}{\partial u^2} + 2\frac{\partial W_2^*}{\partial x} + 2\frac{\partial W_1{}^*}{\partial t} + 3\frac{\partial W_3^*}{\partial u} = 0 \tag{10.79f}$$

$$\frac{\partial W_1^*}{\partial x} = 0 \tag{10.79g}$$

$$\frac{\partial W_2^*}{\partial t} = 0 \tag{10.79h}$$

$$\frac{\partial W_1^*}{\partial u} = 0 \tag{10.79i}$$

$$\frac{\partial W_2^*}{\partial u} = 0 \tag{10.79j}$$

where

$$W^*(t, x, u, p^*, q^*) = W_1^* p^* + W_2^* q^* + W_3^*.$$

It can be seen that $W_1 = W_1(t)$ and $W_2 = W_2(x)$. We seek a nonlinear to linear transformation such that

$$X = \psi \quad \textit{and} \quad T = \tau$$

Therefore,

$$W_1^* = (c_6 - c_4 t) \quad ; \quad W_2^* = (c_5 + c_4 x) \tag{10.80}$$

Using Eqs.(10.80), Eqs.(10.79) become:

$$\frac{\partial W_3^*}{\partial t} + \frac{\partial W_3^*}{\partial x} + \frac{\partial^2 W_3^*}{\partial x \partial t} = 0 \tag{10.81a}$$

$$\frac{\partial^2 W_3^*}{\partial x \partial u} + \frac{\partial W_3^*}{\partial x} = c_4 \tag{10.81b}$$

$$\frac{\partial^2 W_3^*}{\partial u \partial t} + \frac{\partial W_3^*}{\partial t} = -c_4 \tag{10.81c}$$

$$\frac{\partial^2 W_3^*}{\partial u^2} + 3\frac{\partial W_3^*}{\partial u} = 0 \tag{10.81d}$$

Eqs.(10.81b) and (10.81d) give, respectively,

$$\frac{\partial W_3^*}{\partial x} = c_4 + b_1(t,x)e^{-u} \tag{10.82}$$

$$\frac{\partial W_3^*}{\partial t} = -c_4 + b_2(t,x)e^{-u} \tag{10.83}$$

Integrating Eq.(10.82)

$$W_3^* = c_4 x + e^{-u} \int b_1 \, dx + c_8(t) \tag{10.84}$$

Substituting W_3 from Eq.(10.84) into Eq.(10.81d), we get:

$$e^{-u} \int b_1 \, dx - 3e^{-u} \int b_1 \, dx = 0 \tag{10.85}$$

which is true only if $b_1 = 0$. Similarly, it can be shown that $b_2 = 0$ and that

$$c_8(t) = c_7 - c_4 t.$$

Therefore,

$$W_3^*(x,t) = c_7 + c_4(x-t) \tag{10.86}$$

The characteristic function, W^*, can be written as:

$$W^*(t,x,u,p^*,q^*) = (c_6 - c_4 t)p^* + (c_5 + c_4 x)q^* + [c_7 + c_4(x-t)] \tag{10.87}$$

which gives

$$\tau = c_6 - c_4 t \;\; ; \;\; \psi = c_5 + c_4 x \;\; ; \;\; U = c_4(t-x) - c_7 \tag{10.88}$$

The transformation from nonlinear to linear differential equation

$$u = u(v)$$

can be infinitesimally expressed as:

$$U = V\frac{du}{dv} \tag{10.89}$$

Therefore,

$$[c_4(t-x) - c_7] = \left([c_4(t-x) - c_7]v - W_{32}(t,x)\right)\frac{du}{dv}$$

$$(10.90)$$

This requires that $W_{32}(t,x) = 0$, so that

$$u = ln(v) \tag{10.91}$$

is the required nonlinear to linear transformation.

10.4 Limitations of the Inspectional Group Method

While the inspectional method is simple to apply, there are a number of drawbacks that we would like to bring to the attention of the reader:

(a) In the choice of the transformation, there is a certain amount of trial and error and, perhaps, some judgement required. For example, if one were to explore linear groups for the problem treated in section 4.2 , any of the following forms could have been used: (i) $v = c_1 u^{\delta_3} {u_x}^{\delta_4}$ (ii) $v = c_2 u^{\delta_3} {u_t}^{\delta_5}$ (iii) $v = c_3 {u_x}^{\delta_4} {u_t}^{\delta_5}$ (iv) $v = c_4 x^{\delta_1} u^{\delta_3}$
The $\delta' s$ are determined through invariance of the transformation. The validity of the transformation can be ascertained by substituting it into the nonlinear equation and finding out if the required linear form is obtained.

(b) The invariance process under inspectional groups does not take into account the signs between differential terms. For example, both the equations

$$u_{xx} = uu_x + u_t \tag{10.92}$$

and

$$u_{xx} = -uu_x - u_t \tag{10.93}$$

would be invariant under groups (10.15) defined by Eqs.(10.17) and (10.18), giving rise to the same mapping (T). However, only the first equation would transform into the required linear form.

(c) More than a single equation may be invariant under the same group of transformations. One example, as we have seen in (b) is when the signs between the terms of the equation are ignored. The other example is when more terms can be added without affecting the invariance, e.g.,

$$u_{xx} = uu_x + \frac{u^2}{x} + u_t + \frac{u}{t} \tag{10.94}$$

This leads to the question as to what is the most general equation invariant under a given group? Bluman[2] has discussed in his book, a procedure to find the most general equation under an infinitesimal group of transformations. However, the procedure is not applicable when the analysis is sought using inspectional groups.

(d) Consider now the following equations

$$(i)\ u_{xx} = uu_x + u_t \tag{10.92}$$

$$(ii)\, u_{xx} = uu_x + u_{tt} \tag{10.95}$$

Eq.(10.92) is invariant under the group, G_1,

$$G_1 \;:\; [\bar{u} = a^{p} u \;\; ; \;\; \bar{x} = a^{m} x \;\; ; \;\; \bar{t} = a^{2m} t]$$

and Eq.(10.95) is invariant under, G_2,

$$G_2 \;:\; [\bar{u} = a^{p} u \;\; ; \;\; \bar{x} = a^{m} x \;\; ; \;\; \bar{t} = a^{m} t]$$

If we seek an invariant transformation that will transform (10.92) into the heat equation, and (10.95) into the wave equation (eliminate uu_x),i.e.,

$$T: \; u = cv^{\delta_3}(v_x) \tag{10.96}$$

we will get $u = cv_x/v$ for both cases, since the constraints due to the t variable are not brought in. However, only Eq.(10.92) will be linearized.

It can now be seen that although a mapping (T) can be obtained by invoking invariance under an inspectional group of transformations, the required linear equation may not result. The possible reasons have been discussed in this section.

The infinitesimal group method while being cumbersome, circumvents most of the limitations of the inspectional group method. The latter, however, is quick, simple and a useful technique despite its limitations.

10.5 Summary

The concept of invariance has been used to transform a nonlinear differential equation to a linear form. The same concept can be used to relate any two differential equations. Two procedures have been discussed in this chapter; the inspectional group method which is based on simple groups of transformation, and the infinitesimal group method based on infinitesimal Lie groups. While the inspectional group method is easy to apply, it suffers from a number of drawbacks. The infinitesimal group method, although cumbersome, is a mathematically rigorous procedure which overcomes most of the drawbacks of the inspectional group method.

REFERENCES

1. Na, T.Y. and Hansen, A.G.,"Similarity Analysis of Differential Equations by Lie Group",J. of Franklin Institute, Vol.6,p.292 (1972).
2. Bluman, G.W.,"Use of Group Methods for Relating Linear and Nonlinear Partial Differential Equations", Symmetry, similarity and Group Theoretic Methods in Mechanics, Proc. of Symposium, Univ. of Calgary, Canada (1974).

3. Na, T.Y. and Seshadri, R.,"Nonlinear to Linear Dif. Eqs. Using Transformation Groups", 4th Int. Sym. on Large Engineering Systems, The University of Calgary, Canada (1982).
4. Bluman, G.W.,"On the Remarkable Nonlinear Diffusion Equation",J. Math. Phys.,Vol.21,No.5 (1980).
5. Anderson,R.L.,Kumei,S., and Wulfman, C.E.,Rev.Mex. Fis.,Vol.21, No.1, p.35 (1972); Phys. Rev. Lett.,Vol.28, p.988 (1972).
6. Kumei, S. and Bluman, G.W.,"When Nonlinear Differential Equations are Equivalent to Linear Differential Equations",SIAM J. of Appl. Math., Vol.42, No.5 (1982).
7. Na, T.Y. and Seshadri, R.,"From Nonlinear to Linear Equations Using Transformation Groups",Report No. 82-001,Fluid Mechanics Laboratory, University of Michigan, Dearborn Campus (1982).

Chapter 11

MISCELLANEOUS TOPICS

11.1 Reduction of Differential Equations to Algebraic Equations

The reduction of differential equations to algebraic equations developed by Moran and Gaggioli[1], represents yet another application of the group theory. The method was developed based on the notion of Birkhoff[2], that whenever a system of equations transform invariantly under a group of transformations, solutions that are sought are also invariant under the same group.

Consider for the purpose of illustration, the Helmholtz equation:

$$\frac{\partial^2 y}{\partial x_1{}^2} + \frac{\partial^2 y}{\partial x_2{}^2} + \lambda^2 y = 0 \tag{11.1}$$

where λ is a constant.

Let us introduce a two-parameter group defined as

$$G: \ \bar{x}_1 = x_1 + ln(a_1) \ ; \ \bar{x}_2 = x_2 + ln(a_2) \ ; \ \bar{y} = a_1^r a_2^s y \tag{11.2}$$

Under this group of transformations, Eq.(11.1) can be written as:

$$\frac{\partial^2 \bar{y}}{\partial \bar{x}_1{}^2} + \frac{\partial^2 \bar{y}}{\partial \bar{x}_2{}^2} + \lambda^2 \bar{y} = a_1^r a_2^s \left(\frac{\partial^2 y}{\partial x_1{}^2} + \frac{\partial^2 y}{\partial x_2{}^2} + \lambda^2 y\right) = 0 \tag{11.3}$$

Thus, if $y = F(x_1, x_2)$ is any solution of Eq. (11.1), then $\bar{y} = F(\bar{x}_1, \bar{x}_2)$ is a solution of the invariantly transformed equation. Furthermore, if I is a solution to Eq.(11.1) such that $y = I(x_1, x_2)$ transforms under group G into $\bar{y} = I(\bar{x}_1, \bar{x}_2)$, then the invariant solution can implicitly be written as:

$$g(y, x_1, x_2) = K \tag{11.4}$$

where K is a constant. For the group defined by Eq.(11.2), the "invariant variable" can be obtained by eliminating parameters a_1 and a_2. Therefore, the unknown function g in Eq.(11.4) is given by:

$$g(y, x_1, x_2) = ye^{-rx_1 - sx_2} \tag{11.5}$$

Combining Eq.(11.4) and Eq.(11.5), we get:

$$y = Ke^{rx_1 + sx_2} \tag{11.6}$$

Substituting Eq.(11.6) into the Helmholtz equation, Eq.(11.1),with$K \neq 0$, we get the following algebraic equation:

$$r^2 + s^2 + \lambda^2 = 0 \tag{11.7}$$

11.2 Reduction of the Order of an Ordinary Differential Equation

Reduction of the order of ordinary differential equations is well-known in literature(see Ames[3] or Ince[4]). Consider a second order ordinary differential equation:

$$f\left(\frac{d^2y}{dx^2}, \frac{dy}{dx}, y; x\right) = 0 \tag{11.8}$$

Two special classes of Eqs.(11.8) are of particular importance in engineering sciences. In the first class, the independent variable x does not appear in the differential equation, whereas in the second class, the dependent variable y does not appear.

Eq.(11.8) can be rewritten as:

$$f(q, p, y, x) = 0 \tag{11.9}$$

where

$$q = \frac{d^2y}{dx^2} \quad and \quad p = \frac{dy}{dx}$$

We now define an infinitesimal group of transformations:

$$\bar{x} = x + \epsilon\xi(x, y) + O(\epsilon^2) \tag{11.10a}$$

$$\bar{y} = y + \epsilon\theta(x, y) + O(\epsilon^2) \tag{11.10b}$$

$$\bar{p} = p + \epsilon\pi(x, y, p) + O(\epsilon^2) \tag{11.10c}$$

$$\bar{q} = q + \epsilon K(x, y, p, q) + O(\epsilon^2) \tag{11.10d}$$

where ξ, θ, π and K are the infinitesimals. In terms of the characteristic function, W:

$$\xi = \frac{\partial W}{\partial p} \;;\; \theta = p\frac{\partial W}{\partial p} - W$$

$$-\pi = XW \tag{11.11}$$

$$-K = (X^2 + 2qX\frac{\partial}{\partial p} + q^2\frac{\partial^2}{\partial p^2} + q\frac{\partial}{\partial q})\,W$$

where

$$X = \frac{\partial}{\partial x} + p\frac{\partial}{\partial y}$$

The problem is to determine ξ, θ, π and K such that Eq.(11.9) is invariant under the group defined by Eq.(11.10). To this end, we employ the equation

$$Uf = \xi\frac{\partial f}{\partial x} + \theta\frac{\partial f}{\partial y} + \pi\frac{\partial f}{\partial p} + K\frac{\partial f}{\partial q} = 0 \tag{11.12}$$

Substituting Eqs.(11.9) and (11.11) into Eq.(11.12), we can obtain an equation in terms of W.

To get the invariants, the following subsystem is solved:

$$\frac{dx}{\xi} = \frac{dy}{\theta} = \frac{dp}{\pi} = \frac{dq}{K} \tag{11.13}$$

For the case when x is missing, Eq.(11.12) can be written as

$$\theta\frac{\partial f}{\partial y} + \pi\frac{\partial f}{\partial p} + K\frac{\partial f}{\partial q} = 0 \tag{11.14}$$

It can be shown that the characteristic function is given by $W = ap$. The invariants in this case are $I_1 = y$, $I_2 = p$, $I_3 = q$.

Thus, Eq.(11.9) can be expressed in terms of three invariants. If y is taken as the new independent variable, and p as the dependent variable, Eq.(11.8) can be written as

$$F(y, p, \frac{dp}{dy}) = 0 \tag{11.15}$$

Example11.1 One-dimensional Oscillator

Consider now the differential equation for a one-dimensional oscillator with quadratic damping

$$\frac{d^2y}{dt^2} \pm c(\frac{dy}{dt})^2 + f(y) = 0 \tag{11.16}$$

Since the independent variable t is missing from the equation, we can take y and p as the new independent and dependent variables, respectively.Eq.(11.16) is then transformed to

$$p\frac{dp}{dy} \pm cp^2 + f(y) = 0 \tag{11.17}$$

The order of differential equation is seen to be reduced by one.

For the case the dependent variable y in Eq.(11.9) is missing, the characteristic function, W, can be found to be a constant. Thus, the new independent and dependent variables are x and p, and the order of Eq.(11.8) is reduced by one,i.e.,

$$\varphi(x, p, \frac{dp}{dx}) = 0 \tag{11.18}$$

11.3 Transformation from Ordinary to Partial differential Equation-Search for First Integrals

Ames[3] has presented a method by which a search for the first integrals of ordinary differential equations can be carried out. The ordinary differential equation is transformed into a partial differential equation by introducing the functional form of the first integral. Invariance under a group of transformations is then invoked, and the resulting invariant solutions lead to the first integrals of the ordinary differential equation. To introduce the idea, we consider the Lane-Emden equation:

$$\frac{d^2 z}{\partial x^2} + \frac{2}{x}\frac{dz}{dx} + z^k = 0 \tag{11.19}$$

Introducing the transformation $z = y/x$, Eq.(11.19) becomes

$$\frac{d^2 y}{dx^2} + x^{1-k} y^k = 0 \tag{11.20}$$

We now search for a first integral of Eq.(11.20) of the form:

$$F(x, y, y') = C \tag{11.21}$$

Upon differentiation of Eq.(11.21) with respect to x, we get:

$$F_x + y' F_y + y'' F_{y'} = 0 \tag{11.22a}$$

and using Eq.(11,20) to eliminate y", Eq.(11.22a) can be written as

$$F_x + y' F_y - x^{1-k} y^k F_{y'} = 0 \tag{11.22b}$$

Invoking invariance of Eq.(11.22b) under a linear group of transformations:

$$\bar{x} = a^m x \; ; \; \bar{y} = a^n y \; ; \; \bar{y}' = a^r y' \; ; \; \bar{F} = a^p F \tag{11.23a}$$

we find that

$$r = \frac{2m}{1-k} \quad and \quad n = \frac{m(3-k)}{1-k} \qquad (k \neq 1) \tag{11.23b}$$

For $m = 0$, the similarity variables are found by eliminating parameter a from Eq.(11.23) as

$$\xi = \frac{y}{x^{(3-k)/(1-k)}} \; ; \; \eta = \frac{y'}{x^{2/(1-k)}}$$

$$g(\xi, \eta) = \frac{F(x, y, y')}{x^{p/m}} \tag{11.24}$$

Since p is arbitrary, it is equal to zero. Eq.(11.22b) is then transformed to:

$$[(\frac{k-3}{1-k})\xi + \eta]g_\xi - [\frac{2}{1-k}\eta + \xi^k]g_\eta = 0 \tag{11.25}$$

Using the Lagrangian method[8], the following first order differential equation can be obtained:

$$[\frac{2}{1-k}\eta + \xi^k]d\xi + [(\frac{k-3}{1-k})\xi + \eta]d\eta = 0 \tag{11.26}$$

If $w(\xi,\eta)$ is any solution of Eq.(11.26), then

$$g(\xi,\eta) = H[w(\xi,\eta)]$$

is a solution of Eq.(11.26) for any differentiable function H. Since

$$F(x,y,y') = g(\xi,\eta) = H[w(\xi,\eta)] = constant \tag{11.27a}$$

is a first integral of Eq.(11.20), it follows that

$$w(\xi,\eta) = constant = c \tag{11.27b}$$

is an integral of Eq.(11.20). As an example, the first integral for $k = 5$ in Eq.(11.20), takes the form

$$\frac{1}{3}\xi^6 - \xi\eta + \eta^2 = constant$$

or,

$$x(y')^2 - yy' + \frac{1}{3}\frac{y^6}{x^3} = constant$$

11.4 Reduction of Number of Variables by Multiparameter Groups of Transformations

In chapter 3, we had examined methods for reducing the number of independent variables of a partial differential equation by one. If more than one variable needs to be reduced, one may repeat the process of invariance under one-parameter groups. Clearly, this process is tedious. One may then ask whether or not it would be possible to reduce the number of variables by more than one in a single step?

Extending the approach of Morgan, Manohar[9] proposed a method based on two-parameter assumed group of transformations. Invocation of invariance under the group then leads to a series of relationships amongst the constants of the transformation group. Elimination of the parameters of transformations leads to the absolute invariants.

As an application, consider the unsteady, two-dimensional, laminar boundary layer equations:

$$\frac{\partial^2\psi}{\partial y\partial t}+\frac{\partial^2\psi}{\partial x\partial y}\frac{\partial\psi}{\partial y}-\frac{\partial^2\psi}{\partial y^2}\frac{\partial\psi}{\partial x}=\frac{\partial U}{\partial t}+U\frac{\partial U}{\partial x}+\nu\frac{\partial^3\psi}{\partial y^3} \tag{11.28}$$

subject to the boundary conditions:

$$y=0\ :\ \ \frac{\partial\psi}{\partial x}=\frac{\partial\psi}{\partial y}=0 \tag{11.29a}$$

$$y=\infty\ :\ \ \frac{\partial\psi}{\partial y}=U(x,t) \tag{11.29b}$$

Consider first the two-parameter group of transformations defined by

$$G_{21}\ :\ \ t=A^{\alpha_1}\bar{t}\ ;\ \ x=B^{\beta_1}\bar{x}\ ;\ \ y=A^{\alpha_3}B^{\beta_3}\bar{y}$$

$$\psi=A^{\alpha_4}B^{\beta_4}\bar{\psi}\ ;\ \ U=A^{\alpha_5}B^{\beta_5}\bar{U}$$

The condition for the invariance of Eq.(11.28) under this group of transformations requires that

$$\alpha_3=\frac{1}{2}\alpha_1\ ;\ \alpha_4=-\frac{1}{2}\alpha_1\ ;\ \alpha_5=-\alpha_1$$

$$\beta_3=0\ ;\ \beta_4=\beta_5=\beta_1 \tag{11.30}$$

The absolute invariants are therefore

$$\eta=\frac{y}{\sqrt{t}}\ \ ;\ \ f(\eta)=\frac{\psi}{t^{-1/2}x}\ \ ;\ \ c=\frac{U}{t^{-1}x} \tag{11.31a}$$

Similarity solution exists if the mainstream velocity is given by

$$U=c\frac{x}{t} \tag{11.31b}$$

Eq.(11.28) is then transformed to the following equation:

$$f'''+f'+\frac{1}{2}\eta f''-(f')^2+ff''+c(1-c)=0 \tag{11.32a}$$

subject to the transformed boundary conditions:

$$\eta=0\ :\ f=f'=0\ \ ;\ \ \eta=\infty\ :\ f'=1 \tag{11.32b}$$

It must be ensured that the similarity representation constitutes a "complete set of absolute invariants". Theorems relating to the formalism of multiparameter groups are discussed in Moran and Gaggioli[10]. The formalism for elementary cases using the two-parameter "assumed" group is discussed by Ames[11], and can be summarized as follows:

(1) For the two-parameter linear group G_{21}

$$G_{21} : x_1 = A^{\alpha_1}\bar{x}_1 \; ; \; x_2 = B^{\beta_1}\bar{x}_2 \; ; \; x_3 = A^{\alpha_3}B^{\beta_3}\bar{x}_3,$$

$$y_j = A^{\alpha_j}B^{\beta_j}\bar{y}_j \qquad (j = 4, ..., n)$$

The absolute invariants are:

$$\eta = \frac{x_3}{x_1{}^{\alpha_3/\alpha_1} x_2{}^{\beta_3/\beta_1}}$$

and

$$f_j(\eta) = \frac{y_j}{x_1{}^{\alpha_j/\alpha_1} x_2{}^{\beta_j/\beta_1}} \qquad (j = 4, ..., n) \tag{11.33}$$

(2) Similarly, for the group G_{22} defined by

$$G_{22} = x_1 + \alpha_1 A \; ; \; x_2 = B^{\beta_1}\bar{x}_2 \; ; \; x_3 = e^{\alpha_3 A}B^{\beta_3}\bar{x}_3$$

$$y_j = e^{\alpha_j A}B^{\beta_j}\bar{y}_j \qquad (j = 4, ..., n)$$

The absolute invariants are:

$$\eta = \frac{x_3}{e^{(\alpha_3/\alpha_1)x_1} x_2{}^{(\beta_3/\beta_1)}}$$

$$f_j(\eta) = \frac{y_j}{e^{(\alpha_j/\alpha_1)x_1} x_2{}^{(\beta_j/\beta_1)}} \qquad (j = 4, ...n) \tag{11.34}$$

(3) For the group G_{23},

$$G_{23} : x_1 = A^{\alpha_1}\bar{x}_1 \; ; \; x_2 = \bar{x}_2 + \beta_1 B \; ; \; x_3 = A^{\alpha_3}e^{\beta_3 B}\bar{x}_3$$

$$y_j = A^{\alpha_j}e^{\beta_j B}\bar{y}_j \qquad (j = 4, ..., n)$$

The absolute invariants are:

$$\eta = \frac{x_3}{x_1{}^{\alpha_3/\alpha_1} e^{(\beta_3/\beta_1)x_2}}$$

and

$$f_j(\eta) = \frac{y_j}{x_1{}^{\alpha_j/\alpha_1} e^{(\beta_j/\beta_1)x_2}} \qquad (j = 4, ..., n) \tag{11.35}$$

(4) For the group G_{24},

$$G_{24} : x_1 = \bar{x}_1 + \alpha_1 A \; ; \; x_2 = \bar{x}_2 + \beta_1 B \; ; \; x_3 = e^{\alpha_3 A + \beta_3 B}\bar{x}_3$$

$$y_j = e^{\alpha_j A + \beta_j B} \bar{y}_j \qquad (j = 4, ..., n)$$

The absolute invariants are

$$\frac{x_3}{exp(\frac{\alpha_3}{\alpha_1} x_1 + \frac{\beta_3}{\beta_1} x_2)}$$

and

$$f_j(\eta) = \frac{y_j}{exp(\frac{\alpha_j}{\alpha_1} x_1 + \frac{\beta_j}{\beta_1} x_2)} \qquad (j = 4, ..., n) \qquad (11.36)$$

Application of this method has been made to the three- dimensional boundary layer equations[9] and to the nonlinear diffusion[11].

Consider again Eq.(11.28). In an abbreviated form it can be written as:

$$F = p_{333} + \phi - p_{13} - p_3 p_{23} + p_2 p_{33} = 0 \qquad (11.37)$$

where

$$p_1 = \frac{\partial \psi}{\partial t} \ ; \ p_2 = \frac{\partial \psi}{\partial x} \ ; \ p_3 = \frac{\partial \psi}{\partial y} \ ; \ p_{13} = \frac{\partial^2 \psi}{\partial t \partial y} \ ; \ etc$$

and

$$\phi = \frac{\partial U}{\partial t} + U \frac{\partial U}{\partial x}$$

On applying the infinitesimal group method, the differential equation $F = 0$ as given in Eq.(11.37), will be invariant under the two-parameter infinitesimal transformation group:

$$t^* = t + \epsilon_1 \alpha_1(t, x, y, \psi) + \epsilon_2 \bar{\alpha}_1(t, x, y, \psi)$$

$$x^* = x + \epsilon_1 \alpha_2(t, x, y, \psi) + \epsilon_2 \bar{\alpha}_2(t, x, y, \psi)$$

....

....

$$p^*{}_2 = p_2 + \epsilon_1 \pi_2(t, x, y, \psi, p_1, p_2, p_3)$$
$$+ \epsilon_2 \bar{\pi}_2(t, x, y, \psi, p_1, p_2, p_3)$$

....

....

$$p^*{}_{333} = p_{333} + \epsilon_1 \pi_{333}(t, x, y, ..., p_{333})$$
$$+ \epsilon_2 \bar{\pi}_{333}(t, x, y, ..., p_{111})$$

Invariance of Eq.(11.37) gives:

$$c_1 U_1 F + c_2 U_2 F = 0 \qquad (11.38)$$

Since c_1 and c_2 are arbitrary, we have

$$U_1F = 0 \quad and \quad U_2F = 0 \tag{11.39}$$

simultaneously. In their expanded form, they can be written as:

$$\alpha_1 \frac{\partial F}{\partial t} + \alpha_2 \frac{\partial F}{\partial x} + \alpha_3 \frac{\partial F}{\partial y} + \varsigma \frac{\partial F}{\partial \psi} + \pi_i \frac{\partial F}{\partial p_i}$$

$$\pi_{ij} \frac{\partial F}{\partial p_{ij}} + \pi_{ijk} \frac{\partial F}{\partial p_{ijk}} = 0 \tag{11.40a}$$

and

$$\bar{\alpha}_1 \frac{\partial F}{\partial t} + \bar{\alpha}_2 \frac{\partial F}{\partial x} + \bar{\alpha}_3 \frac{\partial F}{\partial y} + \bar{\varsigma} \frac{\partial F}{\partial \psi} + \bar{\pi}_i \frac{\partial F}{\partial p_i}$$

$$+\bar{\pi}_{ij} \frac{\partial F}{\partial p_{ij}} + \bar{\pi}_{ijk} \frac{\partial F}{\partial p_{ijk}} = 0 \tag{11.40b}$$

Putting F from Eq.(11.37) into Eq.(11.40), we get:

$$\pi_{333} + \frac{\partial \phi}{\partial t}\alpha_1 + \frac{\partial \phi}{\partial x}\alpha_2 - \pi_{13} - p_3\pi_{23} - p_{23}\pi_3$$

$$+\pi_2 p_{33} + p_2\pi_{33} = 0 \tag{11.41a}$$

and

$$\bar{\pi}_{333} + \frac{\partial \phi}{\partial t}\bar{\alpha}_1 + \frac{\partial \phi}{\partial x}\bar{\alpha}_2 - \bar{\pi}_{13} - p_3\bar{\pi}_{23} - p_{23}\bar{\pi}_3$$

$$+\bar{\pi}_2 p_{33} + p_2\bar{\pi}_{33} = 0 \tag{11.41b}$$

The next step is to express the transformation functions or the infinitesimals α_1, α_2,etc., in Eq.(11.41) in terms of a characteristic function W; and $\bar{\alpha_1}, \bar{\alpha}_2$, etc., in Eq.(11.41) in terms of a second characteristic function $\bar{W}$. This differs from the one-parameter method in that two characteristic functions, instead of one, have to be determined from Eqs.(11.41a) and (11.41b). We choose, as an example, two groups as follows:

$$G_1 \ : \ W; \ independent \ \ of \ \ p_1$$

$$G_2 \ : \ \bar{W}; \ independent \ \ of \ \ p_2$$

By following the same procedure as in the one-parameter method, the characteristic function W_1 and W_2 can be found from Eqs.(11.41a) and (11.41b). respectively. The results are

$$W_1 = \left[W_{111}x + W_{112}(t) \right] p_2 + \frac{\partial \theta}{\partial x} p_3 - \left[W_{111}\psi + \frac{dW_{112}}{dt} y - \frac{\partial \theta}{\partial t} \right]$$

and

$$W_2 = (W_{211}t + W_{212})p_1 + (\frac{1}{2}W_{211}y + \frac{\partial\bar{\theta}}{\partial x})p_3 + \frac{1}{2}W_{211}\psi + \frac{\partial\bar{\theta}}{\partial t} \tag{11.42}$$

where W_{111}, W_{211} and W_{212} are constants, W_{112} is an arbitrary function of t, and θ and $\bar{\theta}$ are arbitrary functions of t and x. In addition, the following two equations have to be satisfied:

$$\frac{\partial\phi}{\partial x}[W_{111}x + W_{112}(t)] - \phi W_{111} - \frac{d^2 W_{112}}{dt} = 0$$

and

$$\phi W_{211} + \frac{\partial\phi}{\partial t}(W_{211}t + W_{212}) = 0$$

With the characteristic functions W_1 and W_2 for the two groups, G_1 and G_2 known, the next step is to find the absolute invariants. For the combined two-parameter group of transformations, the absolute invariants can be solved from the following system of equations:

$$\frac{dt}{\alpha_1 + a\bar{\alpha}_1} = \frac{dx}{\alpha_2 + a\bar{\alpha}_2} = \frac{dy}{\alpha_3 + a\bar{\alpha}_3}$$

$$= \frac{d\psi}{\varsigma + a\bar{\varsigma}} = c_1\epsilon_1 \tag{11.43}$$

where $a = c_2/c_1$. Substituting the characteristic functions into Eq.(11.43), we get:

$$\frac{dt}{a(W_{211}t + W_{212})} = \frac{dx}{W_{111}x + W_{112}(t)} = \frac{dy}{\theta_x + a(0.5W_{211}y + \bar{\theta}_x)}$$

$$= \frac{d\psi}{(W_{111}\psi + (W_{112})_t y + \theta_t) + a(\bar{\theta}_t - 0.5W_{211}\psi)} = c_1\epsilon_1 \tag{11.44}$$

As an example, consider the case in which $\theta, \bar{\theta}$, W_{112} and W_{212} are all zero. Eq.(11.44) then becomes:

$$\frac{dt}{aW_{211}t} = \frac{dx}{W_{111}x} = \frac{dy}{0.5aW_{211}y} = \frac{d\psi}{W_{111}\psi - 0.5aW_{211}\psi}$$

$$= c_1\epsilon_1 \tag{11.45}$$

The three independent solutions are

$$\frac{x}{t^{W_{111}/(W_{211}a)}} = c_1 \tag{11.46a}$$

$$\frac{y}{t^{1/2}} = c_2 \tag{11.46b}$$

and

$$\frac{\psi}{t^{\left(W_{111}/(aW_{211})-1/2\right)}} = c_3 \tag{11.46c}$$

As a final step, the parameter "a" has to be eliminated from Eq.(11.46). We then get

$$\frac{y}{t^{1/2}} = c_2 \quad and \quad \frac{\psi}{xt^{-1/2}} = \frac{c_3}{c_1} \tag{11.47a, b}$$

which are obtained by eliminating a from Eqs.(11.46a) and (11.46c). The similarity variables are therefore

$$\eta = \frac{y}{t^{1/2}} \quad ; \quad f(\eta) = \frac{\psi}{xt^{-1/2}} \tag{11.48}$$

which are the same as those obtained by Manohar's method. The boundary conditions at the edge of the boundary layer is then transformed to:

$$\eta = \infty : \quad \frac{x}{t} f'(\infty) = U(x,t)$$

which gives:

$$U(x,t) = \frac{x}{t}$$

To show the general nature of this method, consider now two other groups, namely,

$$G_3 \ : \ a \ general \ infinitesimal \ transformation$$

$$G_4 \ : \ a \ general \ infinitesimal \ transformation \quad with \ W_4$$
$$independent \ of \ p_3$$

By following the same steps as for groups G_1 and G_2, we get:

$$W_3 = (W_{311}t + W_{312})p_1 + [W_{321}x + W_{322}(t)]p_2$$
$$+ (\frac{1}{2}W_{311}y + \frac{\partial B}{\partial x})p_3 + (\frac{1}{2}W_{311} - W_{321})\psi$$
$$- \frac{dW_{322}}{dt}y + \frac{\partial B}{\partial t}$$

and

$$W_4 = W_{41}p_1 + [W_{421}x + W_{422}(t)]p_2 - W_{421}\psi - \frac{dW_{422}}{dt}y$$

$$+ W_{4322}(t) \tag{11.49}$$

where $W_{311}, W_{312}, W_{321}, W_{41}$, and W_{421} are constants, W_{322}, W_{422} and W_{4322} are functions of t, and $B(t,x)$ is an arbitrary function of t and x. In addition, the following equations have to satisfied:

$$\frac{3}{2}\phi W_{311} + \frac{\partial \phi}{\partial t}(W_{311} t + W_{312}) + \frac{\partial \phi}{\partial x}[W_{321}(t)x + W_{322}(t)]$$

$$-\phi[\frac{1}{2} W_{311} - W_{321}(t)] - p_3 \frac{d W_{32}}{dt} = 0 \tag{11.50a}$$

and

$$W_{41}\frac{\partial \phi}{\partial t} + [W_{421} x + W_{422}(t)]\frac{\partial \phi}{\partial x} + W_{421}\phi = 0 \tag{11.50b}$$

According to the theorems presented in section 2.8, the absolute invariants can be solved by the following system of equations:

$$\frac{dt}{(W_{311} t + W_{312}) + a W_{41}}$$

$$= \frac{dx}{[W_{321} x + W_{322}(t)] + a[W_{421} x + W_{422}(t)]}$$

$$= \frac{dy}{\frac{1}{2} W_{311} y + \frac{\partial B}{\partial x}}$$

$$= \frac{d\psi}{-[.5 W_{311} - W_{321}]\psi + W_{322t} y - B_t + a[W_{421}\psi + W_{422} y - W_{4322}]}$$

$$= e_1 d\epsilon_1 \tag{11.51}$$

As an example, we consider the case in which B, W_{322}, W_{421} and W_{4322} are all zero and W_{321} and W_{422} are constants. For this case, Eq.(11.51) becomes:

$$\frac{dt}{W_{311} t + W_{312} + a W_{41}} = \frac{dx}{W_{321} x + a W_{432}} = \frac{dy}{\frac{1}{2} W_{311} y}$$

$$= \frac{d\psi}{(-\frac{1}{2} W_{311} + W_{321})\psi} = e_1 d\epsilon_1 \tag{11.52}$$

The three independent solutions to Eq.(11.52) are:

$$\frac{W_{321} x + a W_{422}}{(W_{311} t + W_{312} + a W_{41})^n} = c_1$$

$$\frac{y}{\sqrt{W_{311} t + W_{312} + a W_{41}}} = c_2$$

$$\frac{\psi}{(W_{311}t + W_{312} + aW_{41})^{n-1/2}} = c_3$$

where $n = W_{321}/W_{311}$. For the special case in which n equals to 1 and $W_{312} = 0$,elimination of "a" from the above equations results in:

$$\frac{y}{\sqrt{W_{432}t - W_{41}x}} = c_4 \quad and \quad \frac{\psi}{\sqrt{W_{422}t - W_{41}x}} = c_5$$

The similarity variables are therefore:

$$\eta = \frac{y}{\sqrt{W_{422}t - W_{41}x}} \ ; \ f(\eta) = \frac{\psi}{\sqrt{W_{423}t - W_{41}x}} \tag{11.53}$$

which is the second transformation obtained by Schuh[12].

The boundary conditions at the edge of the boundary layer is transformed to the form:

$$\eta = \infty : \ f(\infty) = U(x,t)$$

which means U has to be a constant. It can be shown that this form of U satisfies Eqs.(11.50). For other values of n, the process of eliminating n is quite difficult. No attempt will be made to discuss this problem here.

11.5 Self-Similar Solutions of the First and Second Kind

Self-similar solutions result from the invocation of invariance under a dimensional or an affine group of transformations. Such solutions can be expressed as:

$$u(x,t) = m(t)F\left[\frac{x}{\ell(t)}\right] \tag{11.54}$$

where x and t are independent variables that may sometimes be interpreted as a spatial coordinates and time, respectively. The term "self-similar" comes from the fact that the spatial distribution of the characteristics of motion,i.e., of the dependent variables, remains similar to itself at all times during the motion.

Self-similar problems have been investigated for some time by Soviet researchers. Sedov[13] has extended the traditional dimensional analysis techniques to obtain self-similar solutions. In his method, self-similar solutions represent solutions of the degenerate problems for which all constant parameters entering the initial and boundary conditions and having the dimensions of the independent variables vanish or become infinite. Barenblatt and Zel'dovich[14] have pointed out that self-similar solutions describe the "intermediate asymptotic" behavior of solutions of wider classes of initial, boundary and mixed problems,i.e., they describe the behavior of these solutions away from the boundaries of the region where in a sense the solution is no longer dependent on the details of the initial and/or boundary conditions, but the system is still far from being in a state of equilibrium.

The treatment of self-similar solutions as intermediate asymptotics enables one to understand the role of dimensional analysis in their construction. The use of dimensional analysis implies a certain regularity of the limiting process when going from the original non-self-similar solution of a non-degenerate problem description to the degenerate self-similar description. Such problems are known as "self-similar problems of the first kind". Sedov[13] provides a number of self-similar solutions of the first kind, especially for problems in gas dynamics. There are, however, a wide class of self-similar problems in which the similarity variables are determined not only from dimensional considerations alone, but from the study of the transition process of the non-self-similar problem to an intermediate asymptotic. This transition is not regular and dimensional considerations must be supplemented by the solution of a certain "eigenvalue problem". These problems are classified as "self-similar problems of the second kind" [14,15]. In terms of group invariance, solutions obtained using traditional dimensional analysis would give rise to self-similar solutions of the first kind, and solutions obtained by the use of affine groups would lead to self-similar solutions of the second kind.

The formation of a blast wave produced by an intense explosion discussed in section 8.2 is an example of self-similar solution of the first kind. The exponent α is equal to 2/5 from dimensional considerations, and the parameter C is determined by satisfying the energy integral, Eq.(8.47). The motion of the shock front,$S(t)$, was described in Eq. (8.52).

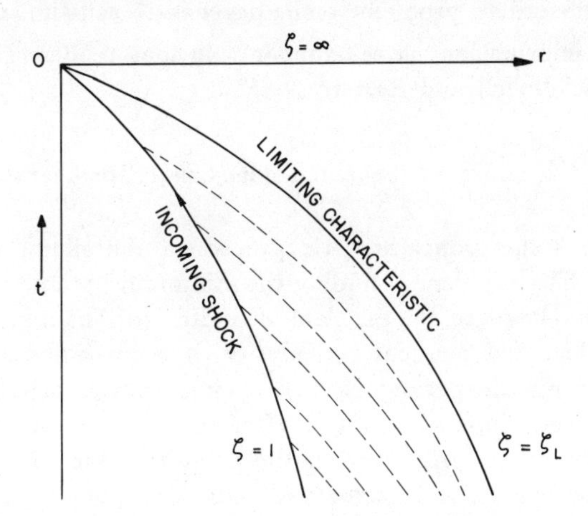

Fig. 11.1 The Implosion Problem

The problem of an implosion of a spherical shock wave is an example of self-similar problem of the second kind (see Fig.11.1). A spherically symmetric shock wave travels to the center of symmetry through a gas of uniform initial density ρ_0 and zero pressure. The origin for time $t = 0$ is taken as the instant of collapse,i.e., when $R(t) = 0$. Therefore, the time up to the instant of collapse is negative. The similarity transformation and representation is given by Eqs.(8.49) and (8.50), respectively. For this problem α cannot be determined by dimensional arguments. The wave could be thought of as being generated by a "spherical piston" which pushes the gas inwards and imparts a certain amount of energy to it. As the wave converges to the center, the energy becomes concentrated at the front and the wave is strengthened. The conservation laws do not hold as was previously the case in the intense explosion problem. Therefore α cannot be determined using the conservation integral. The variation of $R(t)$ with t would no longer be proportional to $t^{2/5}$ but to some exponent α, such that $R(t)$ is proportional to t^{α} .

For the implosion problem, the limiting characteristic through the origin lies in the region of disturbance. Therefore, during the integration of Eqs.(8.50) and (8.51) between $\varsigma = 1$ and $\varsigma = \infty$, the singularity will occur on the curve

$$(V-1)^2 - A^2 = 0$$

However, if one is to expect a solution to continue smoothly across the limiting characteristic, then the right hand side of Eq.(8.50) would be zero. The exponent α is chosen so that the solution is non-singular. The parameter C is obtained from consideration of the limiting passage from the original non-self-similar problem to the degenerate self-similar problem.

For more information on self-similar solutions readers should refer to Sedov[13] and Zel'dovich and Raizer[15].

11.6 Normalized Representation and Dimensional Consideration

In chapter 3 the group-theoretic procedure of Hellums and Churchill was described. The problem of finding the minimum parametric description can be directly related to the problem of finding (a) the minimum description in terms of the independent variables or (b) a nondimensional representation. If the minimum parametric description involves arbitrary reference variables, then the elimination of the reference variables would lead to similarity transformations. However, if the arbitrary reference variables are completely specified, then a normalized representation would result. This type of representation is suitable for scale-modeling and semi-analytical investigations.

We will now consider the problem of forced vibration of a single degree of freedom spring-mass system, as shown in fig.11.2. M is mass, K is the stiffness of the spring,x is displacement and t is the time. The periodic force acting on the mass is $F = F_0 cos(\beta t)$, where F_0 is the amplitude of the periodic force and β is the forcing frequency.

The equation of motion leading to the vibration response is [20]:

$$M\frac{d^2x}{dt^2} + Kx = F_0 cos(\beta t) \tag{11.55}$$

We introduce the transformations

$$\bar{x} = \frac{x}{x_0} \quad and \quad \bar{t} = \frac{t}{t_0}$$

such that $\bar{x}$ and $\bar{t}$ are non-dimensional. The arbitrary reference variables x_0 and t_0 are determined by seeking a minimum parametric description. The transformed equation of motion is

$$\frac{M}{Kt_0^2}\left(\frac{d^2\bar{x}}{d\bar{t}^2}\right) + \bar{x} = \frac{F_0 t_0^2}{Mx_0} cos(\beta t_0 \bar{t}) \tag{11.56}$$

The parametric description of the problem can be written as

$$\bar{x} = \phi(\bar{t}; \frac{M}{kt_0^2}, \frac{F_0 t_0^2}{Mx_0}, \beta t_0) \tag{11.57}$$

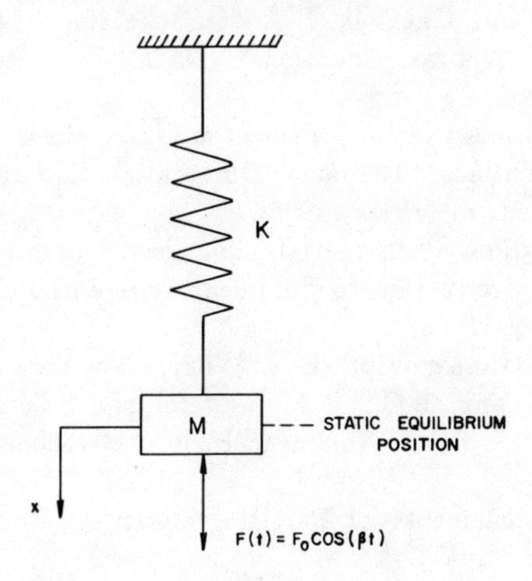

Fig. 11.2 Spring Mass System

If we want to determine the steady state response, we should identify x_0 with the static deflection under dynamic load (F_0),i.e.,

$$\delta_{static} = \frac{F_0}{K}$$

For a minimum parametric description, we set

$$\frac{M}{Kt_0^2} = 1 \quad and \quad \frac{F_0 t_0^2}{Mx_0} = 1$$

Therefore, we get

$$t_0 = \frac{1}{\omega_n} \quad and \quad x_0 = \delta_{static}$$

The resulting problem statement is given by the expression

$$\frac{x}{(F_0/K)} = \phi(\omega_n t; \frac{\beta}{\omega_n}) \tag{11.58}$$

The analytical solution for the steady state response has been determined[20] as:

$$x = \frac{(F_0/K)}{1 - (\frac{\beta}{\omega_n})^2} sin(\omega_n t) \tag{11.59}$$

This example is an illustration of the use of the Hellums- Churchill procedure for obtaining a normalized representation of a boundary value problem.

REFERENCES

1. Moran, M.J. and Gaggioli, R.A.,"On the Reduction of Differential Equations to Algebraic Equations", Math. Res. Rept. 925,Univ. of Wisconsin (1968).
2. Birkhoff,G.,Hydrodynamics, Princeton Univ. Press (1950).
3. Ames,W.F.,Nonlinear Ordinary Differential Equations in Transport Processes, Academic Press (1968).
4. Ince, E.L., Ordinary Differential Equations, Dover (1956).
5. Davis, H.T., Introduction to Nonlinear Differential and Integral Equations, Dover (1962).
6. Stoker, J.J.,Nonlinear Vibrations, Wiley, New York (1950).
7. Na,T.Y. and Tang, S.C.,"A Method for the Solution of Heat Conduction Equation With Nonlinear Heat Generation", ZAMM, Vol.49 (1969).
8. Sneddon, I.N.,Elements of Partial Differential Equations, McGraw-Hill, New York (1957).
9. Manohar, R.,"Some Similarity Solutions for Partial Differential Equations of Boundary Layer Equations", Math. Res. Center, Rept.375, Univ. of Wisconsin (1963).
10. Moran, M.J. and Gaggioli, R.A.,"Similarity for a Real Gas Boundary Layer Flow", Math. Res. Center, Rept. No.925, Univ. of Wisconsin (1968).
11 Ames, W.F.,"Similarity for the Nonlinear Diffusion Equation", I and EC Fundamentals, Vol.4, No.1 (1965).
12. Schuh, H.,"Uber die Ahnlichen Losungen der Instationaren Laminaren Grenzschichtgleichung in inkompressibler Stromung", Fifty Years of Boundary layer Research, Braunschweig (1955).

13. Sedov, L.I., Similarity and Dimensional Methods in Mechanics, Academic Press, New York (1959).
14. Barenblatt, G.I. and Zel'dovich, Ya.B.,"Self-similar Solutions are Intermediate Asymptotics", Ann. Rev. of Fluid Mech. (1972).
15. Zel'dovich, Ya.B. and Raizer, Yu. P., Physics of Shock Waves and High Temperature Hydrodynamic Phenomena, (translation), Vol.2, Academic Press (1967).
16. Butler, D.S.,"Converging Spherical and Cylindrical Shocks", Rept. No.54, Armament Research Development Establishments, Ministry of Supply, Port Halstead, Kent (1954).
17. Sakurai, A.,"On the Problem of a Shock Wave Arriving at the Edge of a Gas", Comm. Pure Appl. Math.,13, pp.353-370 (1960).
18. Hayes, W.D.,"Self-Similar Strong Shocks in an Exponential Medium", J. Fluid Mech.,32, pp.305-315 (1968).
19. Whitham, G.B., Linear and Nonlinear Waves, Wiley (1974).
20. Thomson,W.T.,Theory of Vibrations and Applications, Prentice-Hall (1972).
21. Baker, W.E., Westine, P.S. and Dodge, F.T., Similarity Methods in Engineering Dynamics, Spartan Books (1973).

Problems

1. Show that the transformations

$$\bar{x} = A^{\alpha_1} x \quad ; \quad \bar{y} = A^{\alpha_2} y$$

forms a group. What is the identity element?

2. Show that the infinitesimal transformations

$$\bar{x} = x + \epsilon X(x, y) + O(\epsilon^2)$$

$$\bar{y} = y + \epsilon Y(x, y) + O(\epsilon^2)$$

form a group. What is the identity for the transformation? By evaluating the Jacobian of the transformation, show that the inverse exists.

3. Are the parameters in the group

$$\bar{x} = x + (a_1 + a_2)$$

$$\bar{y} = y + a_1$$

essential or non-essential?

4. Find the infinitesimal group of transformations corresponding to the finite group of transformations

$$\bar{x} = e^a x + e^a (1 - e^a) y$$

$$\bar{y} = e^{2a} y$$

5. Find the finite group of transformations corresponding to the infinitesimal group

$$\bar{x} = x + \epsilon(x - y) + O(\epsilon^2)$$

$$\bar{y} = y + \epsilon(2y) + O(\epsilon^2)$$

6. Find the invariant functions of the following:

$$(i)\ Uf = x^2 \frac{\partial f}{\partial x} + y \frac{\partial f}{\partial y}$$

$$(ii)\ Uf = x \frac{\partial f}{\partial x} - y \frac{\partial f}{\partial y}$$

7. Find the global groups corresponding to the following:

$$(i)\ Uf = x^2 \frac{\partial f}{\partial x} + xy \frac{\partial f}{\partial y}$$

$$(ii)\ Uf = e^{x}\frac{\partial f}{\partial x}$$

$$(iii)\ Uf = x\frac{\partial f}{\partial x}$$

8. What is the significance of the characteristic function? Comment on the differences between the Na-Hansen characteristic function method and the Bluman-Cole infinitesimal group method of invariance analysis.
9. When does the method of traditional dimensional analysis fail to yield similarity transformations? What is the effect of assigning independent dimensions to different directions of the coordinates? Comment on the consequences of specifying too many independent dimensions for a physical problem.
10. The classical separation of variables solution to the linear heat equation

$$\frac{\partial^2 u}{\partial x^2} = \frac{\partial u}{\partial t}$$

or the linear wave equation

$$\frac{\partial^2 u}{\partial x^2} = \frac{\partial^2 u}{\partial t^2}$$

can be written as

$$u(x,t) = X(x)\,T(t)$$

Invoking invariance under the inspectional group of transformations,

$$G\ :\ (\bar{x} = x\ ;\ \ \bar{t} = t + \beta_1 a\ ;\ \ \bar{u} = e^{\beta_2 a} u$$

show that the separation of variables solution can be obtained as:

$$u(x,t) = e^{\beta_2 t/\beta_1} H(\varsigma)\ \ ;\ \ \varsigma = x$$

For certain forms of nonlinear partial differential equations, can a separation of variables type of solution be expected?
11. Find the fundamental solution for the linear wave equation

$$u_{tt} - u_{xx} = \delta(x)\delta(t)$$

subject to the initial conditions

$$u(x,0) = u_t(x,0) = 0$$

by using the inspectional and infinitesimal group methods.

12. Consider the nonlinear Klein-Gordon equation

$$\varphi_{tt} - \varphi_{xx} + V'(\varphi) = 0$$

Invoking invariance under a group of transformation, the invariant solution can be obtained as

$$\varphi = \psi(\xi) \ ; \ \xi = kx - \omega t$$

Solutions such as above arises in connection with dispersive waves. φ is the frequency and k is the wave number, where $\omega = \omega(k)$. The resulting ordinary differential equation takes the form

$$(\omega^2 - k^2)\psi_{\xi\xi} + V'(\psi) = 0$$

Show that the solution to the above equation can be written as

$$\xi = \left[\frac{1}{2}(\omega^2 - k^2)\right]^{1/2} \int \frac{d\psi}{[A - V(\psi)]^{1/2}}$$

where A is a constant of integration. Periodic solutions are obtained when ψ oscillates between two simple zereos of $[A - V(\psi)]$. If the zeros are denoted by ψ_1 and ψ_2 such that

$$\psi_1 < \psi < \psi_2 \ , \ A - V(\psi) \geq 0 \ , \ \omega^2 - k^2 > 0,$$

then with the period in θ normalized to 2π, the periodic solution written as

$$2\pi = \left[\frac{1}{2}(\omega^2 - k^2)\right]^{1/2} \int \frac{d\psi}{[A - V(\psi)]^{1/2}}$$

Find the solutions $\phi = \psi(\xi)$ and the dispersion relationship for the following forms of V(ψ):

$$(i) \qquad V(\psi) = \frac{1}{2}\psi^2$$

$$(ii) \qquad V(\psi) = \frac{1}{2}\psi^2 + \alpha\psi^4$$

$$(iii) \qquad V(\psi) = \beta\psi^3.$$

Details of the analysis of dispersive waves are given in a book by G.B. Whitham entitled "Linear and Nonlinear Waves", Wiley-Interscience, 1974. Involking invariance under an infinitesimal group of transformations, obtain all possible invariant solutions for the Klein-Gordon equation along with corresponding forms of V(ϕ).

13. The differential equations governing free convection on a vertical plate can be written as:

$$\frac{\partial u}{\partial x} + \frac{\partial v}{\partial y} = 0$$

$$u\frac{\partial u}{\partial x} + v\frac{\partial u}{\partial y} = \nu\frac{\partial^2 u}{\partial y^2} + g\frac{\theta_w - \theta_\infty}{\theta_\infty}\theta$$

$$u\frac{\partial \theta}{\partial x} + v\frac{\partial \theta}{\partial y} = \alpha\frac{\partial^2 \theta}{\partial y^2}$$

where

$$\theta = \frac{\theta - \theta_\infty}{\theta_w - \theta_\infty}.$$

and the boundary conditions are

$$y = 0 : \quad u = v = 0 \; ; \; \theta = 1$$

$$y = \infty : \quad u = 0 \; ; \; \theta = 0$$

Using the modified dimensional procedure, show that the three π terms are

$$\pi_1 = \frac{\nu}{\alpha} \; ; \; \pi_2 = \frac{g(\theta_w - \theta_\infty)}{\nu^2\theta_\infty}\frac{y^4}{x}$$

$$\pi_3 = u\left[\frac{\theta_\infty}{g(\theta_w - \theta_\infty)x}\right]^{1/2}$$

Show that the similarity transformation is given by

$$u = \left[\frac{g(\theta_w - \theta_\infty)x}{\theta_\infty}\right]^{1/2} f(\pi_1, \pi_2)$$

and

$$v = \sqrt{\nu}\left[\frac{g(\theta_w - \theta_\infty)}{\theta_\infty}\right]^{1/4} x^{-1/4} g(\pi_1, \pi_2)$$

For a wall temperature distribution of the form

$$\theta_w - \theta_\infty = Bx^n$$

and with blowing through the surface, show that a similarity solution is possible if v at the wall is proportional to $x^{(n-1)/4}$.

14. The equations for the problem of velocity impact of a semi-infinite one-dimensional rod can be written as

$$\frac{\partial \sigma}{\partial x} = -\rho\frac{\partial v}{\partial t}$$

$$\frac{\partial e}{\partial t} = -\frac{\partial v}{\partial x}$$

$$\frac{\partial e}{\partial t} = k\sigma^p e^q$$

x is the Lagrangian space coordinate, t is the time, σ, e and v are the nominal compressive stress, nominal compressive strain and particle velocity, respectively. k, p and q are material constants for rate-sensitive strain-hardening materials. The boundary conditions are

$$v(0,t) = v_c t^\alpha \quad , \quad t > 0$$

$$v(x,0) = \sigma(x,0) = e(x,0) = 0 \ , \quad x \geq 0$$

Based on physical considerations,

$$v(\infty,t) = \sigma(\infty,t) = e(0,\infty) = 0$$

Using the Hellums-Churchill or Birkhoff- Morgan method show that the similarity transformation can be written as:

$$v(x,t) = v_c t^\alpha f(\xi)$$

$$\sigma(x,t) = \rho^{\frac{q-1}{q-p-1}} v_c^{\frac{2(q-1)}{q-p-1}} k^{\frac{1}{q-p-1}} t^{\frac{2\alpha(q-1)+1}{q-p-1}} g(\xi)$$

$$e(x,t) = \rho^{\frac{p}{p-q+1}} v_c^{\frac{2q}{p-q+1}} k^{\frac{1}{p-q+1}} t^{\frac{2\alpha p+1}{p-q+1}} h(\xi)$$

where

$$\xi = x[k^{\frac{1}{q-p-1}} \rho^{\frac{p}{q-p-1}} v_c^{\frac{p+q-1}{q-p-1}} t^{\frac{p-q-\alpha(p+q-1)}{p-q+1}} .$$

(a) Obtain the transformed ordinary differential equations.
(b) For $p = 1, q = 0, \alpha \neq 0$, show that the solution can be written as

$$v(x,t) = v_c t^\alpha \, 2^{2\alpha} \, \Gamma(\alpha+1) i^{2\alpha} \, erfc(\frac{\xi}{2})$$

where $i^{2\alpha} \, erfc(\xi/2)$ is the repeated integral of the error function.
(c) For $q = 0$ (corresponding to rate-sensitive material) and $\alpha = 0$, show that the solution can be expressed as follows:

$$v(x,t) = v_c \left[1 - \int_0^\xi \frac{d\lambda}{(C + \beta\lambda^2)^{p/(p-1)}}\right]$$

where $\beta = p(p-1)/[2(p+1)]$; and

$$C = \left[\frac{4\beta}{\pi} \frac{\Gamma(p/(p-1)}{\Gamma[p/(p-1) - 1/2]}\right]^{\frac{1}{1-2p/(p-1)}}$$

Reference: Seshadri, R. and Singh, M.C., "Similarity Analysis of Rods of Nonlinear Rate- sensitive Strain-hardening Materials", Archives of Mechanics, 28,1,pp.63-74 (1976).

15. The problem of elastic expulsion of Newtonian fluids from long pressurized tubes can be described by the nonlinear equation

$$\frac{16\mu R_0}{k}\left(\frac{\partial R}{\partial t}\right) = R^3\left(\frac{\partial^2 R}{\partial x^2}\right) + 4R^2\left(\frac{\partial R}{\partial x}\right)^2$$

where R is the inner radius of the tube; the subscript 0 refers to the relaxed condition, and subscript 1 refers to the fully extended condition, x is the axial distance from the point of severance of the tube, t is the time, μ is the fluid viscosity and k is the tube wall compliance. The initial and boundary conditions are

$$R(x,0) = R_1 \quad ; \quad R(0,t) = R_0 \quad ; \quad R(\infty,t) = R_1$$

By the use of any of the techniques described in Chapter 3, show that the similarity transformation can be expressed as

$$\frac{R - R_0}{R_1 - R_0} = \phi(\xi), \quad where \quad \xi = \frac{2x}{R_0}\sqrt{\frac{\mu}{kt}}$$

and the similarity representation can be written as:

$$[\alpha + (1-\alpha)\phi]^3\phi'' + 4(1-\alpha)[\alpha + (1-\alpha)\phi]^2(\phi')^2 + 2\alpha^3\xi\phi' = 0$$

where $\alpha = \frac{R_0}{R_1}$. The auxiliary conditions then become

$$\phi(0) = 0 \quad ; \quad \phi(\infty) = 1$$

The solution of the above ordinary differential equation can be obtained in the form of a quadrature. When R_0 approaches R, i.e., α approaches 1 show that the solution is

$$\phi(\varsigma) = erf(\varsigma).$$

The above problem formulation has applications in biophysics. Specifically, the problem deals with the deformation of tube walls under variations of internal pressure (latex tubes of trees, nerve fibres and blood vessels). Details of the solution are discussed by G.S.H. Lock in a paper entitled "Elastic Expulsion From a Long Tube", Bull. of Math. Biophysics, Vol.31 (1969).

16. The process in which a wetting fluid displaces a non-wetting fluid that initially saturates a porous medium by capillary forces alone is known as imbibition can be found in the area of hydrogeology and petroleum recovery. For a cylindrical piece of a porous matrix under simplifying assumptions, a variable S which depends on the displacing

phase saturation, S_i (such that $S = 1 - \alpha S_i$ and $\alpha = constant$), the imbibition process can be described by the nonlinear equation

$$\frac{\partial S}{\partial t} - c_0\left(\frac{\partial S}{\partial x}\right)^2 + c_1\left(\frac{\partial S}{\partial x}\right) + c_2 S\left(\frac{\partial^2 S}{\partial x^2}\right) + c_3 S = 0$$

c_0, c_1, c_2 and c_3 are constants which can be expressed in terms of the physical parameters of the problem, x is the space coordinate and t is the time. Using Birkhoff-Morgan method, show that the similarity transformation can be expressed as:

$$S(x,t) = t^{-\delta/(\delta+1)} F(\xi), \quad where \ \xi = \frac{x}{t^{1/(2\delta+2)}}$$

Transformation of the partial differential equation into an ordinary differential equation and subsequent integration is discussed by Verma, A.P. and Mishra, S.K. in a paper,"Similarity Solution for Imbibition in Porous Media", which appears in Symmetry, Similarity and Group Theoretic Methods in Mechanics, Calgary, Canada (1974).

17. A nonlinear wave equation, that is encountered in affine connection field theory and also in one-dimensional gas dynamics with a particular local rate of combustion, can be written as:

$$\frac{\partial^2 \phi}{\partial t^2} + 2\phi\frac{\partial \phi}{\partial t} - \frac{\partial^2 \phi}{\partial x^2} = 0$$

$$\phi(x,0) = \alpha(x) \ ; \ \frac{\partial \phi}{\partial t}(x,0) = \beta(x)$$

(a) Using inspectional group procedures, obtain the similarity transformation

$$\phi(x,t) = \frac{1}{t}f(\xi) \ ; \ \xi = \frac{x}{t}$$

What boundary conditions are compatible with the above similarity representation? Integrate the ordinary differential equation analytically or numerically.

(b) Show that invariance of the equation under a group of translations $\bar{x} = x + x_0, \bar{t} = t + t_0, \bar{u} = u$ gives rise to the traveling wave solution

$$\phi = c_3 g(\xi) \ ; \ \varsigma = \frac{c_2 x - c_3 t}{[(c_2/c_3)^2 - 1]}$$

The ordinary differential equation obtained is

$$g'' + 2g\, g' = 0$$

The solution can be written as

$$g = Tanh(\varsigma + \varsigma_0)$$

which is an exact simple wave solution. The initial waveform propagates without change of shape or amplitude at constant velocity (c_3/c_2) in the x direction.
(c) Using a deductive group procedure, obtain various invariant solutions including those described above in (b) and (c).
(d) Is it possible to transform the nonlinear wave equation to any of the following linear forms using inspectional methods?

$$(a) \qquad \frac{\partial^2\theta}{\partial t^2} - \frac{\partial^2\theta}{\partial x^2} = 0$$

$$(b) \qquad \frac{\partial^2\theta}{\partial x \partial t} = 0$$

$$(c) \qquad \frac{\partial^2\theta}{\partial t^2} + \frac{\partial\theta}{\partial t} - \frac{\partial^2\theta}{\partial x^2} = 0$$

Details of the similarity analysis is given in a paper by G. Rosen entitled "Solutions of a certain Nonlinear Wave Equation" that appear in the Journal of Math. Phys., 45, 235 (1966).

18. Using the infinitesimal group of transformations

$$\bar{u} = u + \epsilon U(x, t, u) + O(\epsilon^2)$$

$$\bar{t} = t + \epsilon T(x, t, u) + O(\epsilon^2)$$

$$\bar{x} = x + \epsilon X(x, t, u) + O(\epsilon^2)$$

obtain the invariant solutions for the nonlinear heat conduction equatioin

$$\frac{\partial}{\partial x}[C(u)\frac{\partial u}{\partial x}] = \frac{\partial u}{\partial t}$$

Specifically obtain the following infinitesimals and the associated similarity representation:
(a) $X(x) = Bx + \gamma$
(b) $T(t) = 2A + 2Bt$; $U(u) = 0$; $C(u)$ *is arbitrary*
(b) $X(x) = (\beta + B)x + \gamma$; $T(t) = 2A + 2Bt$
$U(u) = \frac{2\beta}{\nu}(u + k)$; $C(u) = \lambda(u + k)^{\nu}$
(c) $X(x) = (\beta + B)x + \alpha x^2 + \gamma$
$T(t) = 2A + 2Bt$; $U(u) = -\frac{3}{2}(u + k)(2\alpha x + \beta)$
$C(u) = \lambda(u + k)^{-4/3}$
The problem has been worked out in detail by G. W. Bluman in his Ph.D. thesis entitled "Construction of Solutions to Partial Differential Equations by the Use of Transformation Groups", California Institute of Technology, Pasadena, California (1967).

19. The Stokes second problem described in section 6.1 of this book involves the determination of the following boundary value problem:

$$\nu\frac{\partial^2 u}{\partial y^2} = \frac{\partial u}{\partial t}$$

$$u(0,t) = U_0 \ cos(\omega t) \ \ ; \ \ u(\infty,t) = u(y,0) = 0$$

Using the method of pseudo-similarity transformations, obtain the solution to the above boundary value problem. Compare the solution so obtained with the closed form result,

$$y(y,t) = U_0 \ exp - \sqrt{2\omega t}\varsigma \ cos(\omega t - \sqrt{2\omega t}\varsigma)$$

where

$$\varsigma = \frac{y}{2\sqrt{\nu t}}.$$

20. Paul Chambre, in an article entitled "The Laminar Boundary Layer With Distributed Heat Source or Sink" in Applied Scientific Research, Sec. A, Vol.6 discusses the problemof temperature distribution during flow over a flat plate. In a particular case of a distributed heat source, the following special equation for a temperature T is obtained as

$$\frac{1}{Pr}\frac{\partial^2 T}{\partial \varsigma^2} + \frac{1}{2}F\frac{\partial T}{\partial \varsigma} - xF'\frac{\partial T}{\partial x} + \sum_{n=1}^{\infty} nb_n\left(\frac{x}{L}\right)^n = 0$$

The similarity variable is given by

$$\varsigma = \left(\frac{V_0}{\nu}\right)\left(\frac{y}{x^{1/2}}\right);$$

V_0 is the mainstream velocity and $F(\varsigma)$ is the Blasius function. The boundary conditions are

$$T = \sum_{n=0}^{\infty} b_n\left(\frac{x}{L}\right)^n \qquad for \quad \varsigma = \infty$$

$$T = 0 \qquad for \quad \varsigma = 0.$$

Assuming a solution for T of the form

$$T(\varsigma, x) = \sum_{n=0}^{\infty} b_n\left(\frac{x}{L}\right)^n G_n(\varsigma)$$

where $G_n(\varsigma)$ satisfies the equation

$$\frac{1}{Pr}G''_n + \frac{1}{2}FG'_n - nF'G_n + n = 0$$

$$G_n(0) = 0 \ \ ; \ \ G_n(\infty) = 0$$

Systematically derive the steps outlined above using a one-parameter group of transformations.

21. The equation governing heat conduction in spherically symmetric description can be written as

$$\frac{1}{r^2}\frac{\partial}{\partial r}[r^2 k(T)\frac{\partial T}{\partial r}] = \frac{\partial T}{\partial t}$$

If at time $t = 0$, an energy of E is released at $r = 0$, then the law of conservation of energy gives

$$\int_0^\infty T(r,t) 4\pi r^2 \, dr = \frac{E}{\rho c_v} = Q.$$

Assuming $k(T) = \alpha T^n$, obtain the similarity transformation

$$T(r,t) = [\frac{Q^{2/3}}{\alpha t}]^{3/(3n+2)} \phi(\xi), \quad \textit{where} \ \xi = \frac{r}{(\alpha Q^n t)^{1/(3n+2)}}$$

Show that the moving boundary propagates at a finite speed given by

$$R(t) = \xi_0 (\alpha Q^n t)^{1/(3n+2)}$$

where

$$\xi_0 = [\frac{3n+2}{2^{n-1}\pi n} \frac{\Gamma(\frac{5}{2}+\frac{1}{n})}{\Gamma(1+\frac{1}{n})\Gamma(\frac{3}{2})}]^{n/(3n+3)}$$

What happens to the speed of propagation when n approaches to 0? What does this mean from a physical standpoint?

22. In a paper entitled "On Similarity Solutions of Wave Propagation for a General Class of Nonlinear Dissipative Materials" that appeared in the Int'l J. of Nonlinear Mechanics, Vol.11, 1976, Chand et al use the deductive group method based on finite group of transformations to find a number of different invariant solutions for the system of equations:

$$\rho_0 \frac{\partial v}{\partial t} = \frac{\partial \sigma}{\partial x} \qquad (\textit{momentum})$$

$$\frac{\partial v}{\partial x} = \frac{\partial \epsilon}{\partial t} \qquad (\textit{compatibility})$$

$$\sigma + A_1 \epsilon + A_2 (\frac{\partial \epsilon}{\partial t})^m + A_4 (\frac{\partial \sigma}{\partial t})^p = 0$$

$$(\textit{constitutive relationship})$$

An exhaust analysis is presented in the paper, and the thrust of the work is mathematically motivated. Introducing the conditions at the moving boundary for characteristic propagation, examine the problem from a physical standpoint by invoking the ideas presented in chapter 8 of this book.

23. The steady-state temperature distribution for heat conduction with heat generation according to the exponential variation can be written as

$$\frac{1}{r^i}\frac{d}{dr}(r^i\frac{\partial T}{\partial r}) + \beta e^T = 0 \qquad 0 < r < R$$

where $i = 1$ for cylindrical symmetry and $i = 2$ for spherical symmetry. The boundary conditions are given by

$$\frac{dT}{dr} = 0 \quad ; \quad T(1) = 0$$

Transform the two-point boundary value problem into the following initial value problem:

$$\frac{1}{x^i}\frac{d}{dx}(x^i\frac{dy}{dx}) + \beta e^y = 0$$

$$\frac{dy(0)}{dx} = 0 \quad ; \quad y(0) = 0$$

using the group of transformations

$$r = e^{\alpha_1 A}x \quad ; \quad T = y + \alpha_2 A$$

For details of the solution, readers should refer to the paper by Na,T.Y. and Tang, S. C.,- "A Method For the Solution of Conduction Heat Transfer With Nonlinear Heat Generation", ZAMM, 49 (1969).

24. Using the infinitesimal group method described in chapter 10, we will derive the mapping that will transform the Burger's equation

$$v_{xx} = vv_x + v_t$$

into the linear heat equation

$$u_{xx} = u_t$$

The infinitesimals for the heat equation have already been derived in chapter 3. Assume the mapping to be of a general form

$$v = F(x, t, u, u_x, u_t)$$

so that, infinitesimally,

$$V = X\frac{\partial F}{\partial x} + T\frac{\partial F}{\partial t} + U\frac{\partial F}{\partial u} + U_x\frac{\partial F}{\partial u_x} + U_t\frac{\partial F}{\partial u_t}$$

where V_x and V_t are the "extended infinitesimals" of the group for the linear heat equation. For non-zero value of the parameters of the group, a system of partial differential equations can be solved to give the Hopf-Cole transformation

$$v = -2\left(\frac{u_x}{u}\right).$$

Details are given in the article by G.W. Bluman: "Use of Group Methods for Relating Linear and Nonlinear Partial Differential Equations", Symmetry, Similarity and Group Theoretic Methods in Mechanics, Calgary, Canada, 1974.

25. Using the infinitesimal group of transformations, show that the mapping that will transform the equation

$$v_{xx} = v_t + \frac{1}{2}v_x^2$$

into equation

$$u_{xx} = u_t$$

can be obtained as $v = -2ln(u)$.

26. Show that the mapping that will transform equation

$$Z_{xy} - e^Z = 0$$

into the linear form $W_{xy} = 0$, can be derived as

$$Z = ln\left|\frac{2W_x W_y}{W^2}\right|.$$

INDEX